THE STARRY UNIVERSE
The Cecilia Payne-Gaposchkin Centenary

Proceedings of a symposium held at
THE HARVARD-SMITHSONIAN CENTER FOR ASTROPHYSICS
Cambridge, Massachusetts, October 26-27, 2000

EDITED BY

A. G. Davis Philip
Institute for Space Observations &
Union College

Rebecca A. Koopmann
Union College

Contributions of the Institute for Space Observations No. 19

L. Davis Press
Schenectady, New York
USA

20th ANNIVERSARY YEAR

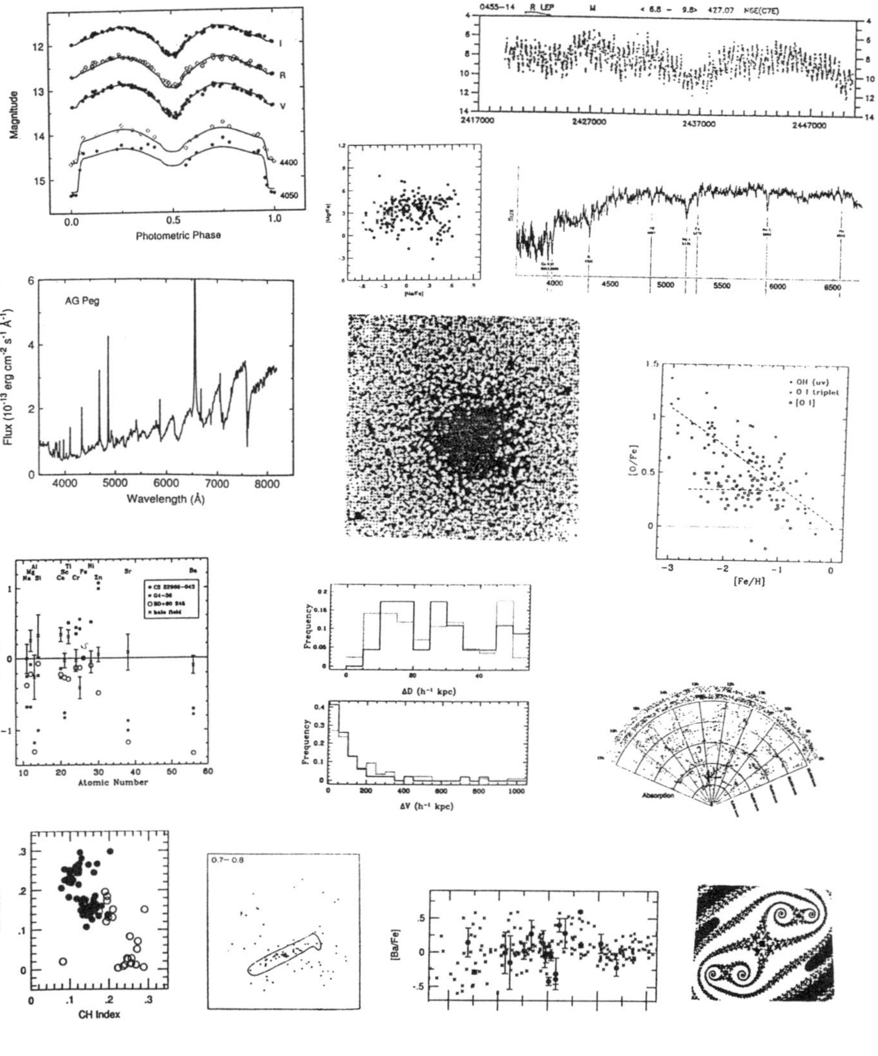

Magnitude
Photometric Phase
0455-14 R LEP M
2417000 2427000 2437000 2447000
[Mg/Fe]
[Na/Fe]
flux
AG Peg
Flux (10^-13 erg cm^-2 s^-1 Å^-1)
Wavelength (Å)
OH (uv)
O I triplet
[O I]
[O/Fe]
[Fe/H]
Δ[X/Fe]
Atomic Number
CS 22966-043
G4-36
BD+80 245
halo field
Frequency
ΔD (h^-1 kpc)
ΔV (h^-1 kpc)
Absorption
CN Index
CH Index
0.7-0.8
[Ba/Fe]

TABLE OF CONTENTS

PREFACE

On October 26 - 27 a meeting was held at the Harvard-Smithsonian Astrophysical Observatory in Cambridge, Massachusetts to honor the centenary of Cecilia Payne-Gaposchkin. Owen Gingerich contacted the L. Davis Press and asked if we would be interested in publishing the proceedings of the meeting. We sent a positive answer and joined the effort to organize and then prepare the proceedings for publication.

The Scientific Organizing Committee was formed of the following people:

Owen Gingerich	Harvard-Smithsonian CfA	Chairman
Andrea Dupree	Smithsonian Astrophysical Observatory	
Katherine Haramundanis	Compaq Computer Corp.	
David Latham	Harvard-Smithsonian CfA	
A. G. Davis Philip	ISO & Union College	
Virginia Trimble	University of California, Irvine.	

Authors of papers sent manuscripts to Schenectady in Word Perfect and in LaTex format. The papers were set up in the correct L. Davis Press format and then checked for errors in the text. The articles were printed out and returned to authors for proofreading. Authors corrected their papers and mailed them back. This cooperation of the authors is vital to the preparation of the final manuscript. The more eyes search papers for errors the better the final product becomes. Owen Gingerich sent in details concerning the list of participants, session chairs, the Observatory Pinafore (to be found in the Appendix) and the Foreword. He also arranged to have color copies of the symposium poster made for insertion into each copy of the book. Kathy Haramundanis helped in preparing the tracing of Cecilia Payne-Gaposchkin which appears on the cover. Janet Mattei sent in the photo of the participants, taken by Ronald Royer.

This is the twentieth year of the L. Davis Press. The first volume, "X-Ray Symposium - 1981" was made using a program called Scripsit on a TRS-80 and printed on a dot-matrix printer. Later I used Final Word on an IBM PC and then various editions of Word Perfect. Now the astronomical world has mostly moved on to LaTex and a Linux environment. A new system, XML, is on the horizon so at some point we will have to learn yet another system for producing papers. A short history of the press can be found in "Desktop Publishing", published by World Scientific in 1992 and edited by Andre Heck.

In the Subject Index there is an area listed under the heading, CPG, where details related to her life and activities have been collected together. Other topics with many entries which are collected together are Figures, Institutions and Places. Pictures of people and places are marked by an asterisk immediately after the page number.

A. G. Davis Philip

FOREWARD

The history of Harvard College Observatory is unusual in the number of talented women who made distinctive and memorable astronomical contributions. Williamina Fleming, Annie Jump Cannon and Antonia Maury were notable pioneers in the spectral classification of stars. Henrietta Leavitt is especially remembered for the period-luminosity relation of Cepheid variables.

But the star among them was Cecilia Payne-Gaposchkin, the astrophysicist who demonstrated that despite the variety of the stellar spectral types, there is a surprising homogeneity of chemical composition among the stars. Her wide-ranging researches included establishing the modern categories of variable stars, the evolutionary placement of variable stars within the Milky Way, and the dependence of Cepheid properties on their location the Large Magellanic Cloud and therefore their evolutionary sweep. With special courage and prescience she entitled her 1952 book "Stars in the Making".

Born in England in 1900, Cecilia Payne came to America in 1923, earning the first astronomy Ph.D. at Harvard College Observatory, in 1925. She then spent her entire astronomical career at Harvard. Cecilia Payne-Gaposchkin became the first woman advanced to the rank of full professor in the Harvard Faculty of Arts and Sciences and at the same time the first woman department chairman. She was the first recipient of the American Astronomical Society's Annie J. Cannon Prize and the first woman to receive their highest honor, the Henry Norris Russell Lectureship. She lived until 1981, just before her eightieth birthday.

As the University Marshall, Richard Hunt, pointed out in his remarks at the symposium dinner, Cecilia Payne-Gaposchkin was not the first tenured woman professor at Harvard, but she was the first woman professor to have received tenure at Harvard. The basis of this curious distinction is that she was tenured by the Harvard Corporation in 1938 as Phillips Astronomer, but she did not become a professor until 1956; in the meantime Sirarpie Der Nersessian, in Byzantine art "at Harvard's Dumbarton Oaks branch in Washington D. C.", Helen Maud Cam, in English History and Cora Dubois, in Anthropology, had become professors. However, when Cecilia Payne-Gaposchkin became Phillips Professor, she was simultaneously appointed department chairman, the first woman to achieve this distinction at Harvard.

At the dinner Richard Hunt also recognized Professor Dudley Herschbach, who announced that he and his wife Georgene, in the interests of equal opportunities for portraiture, were sponsoring a new oil portrait of Cecilia Payne-Gaposchkin for the Faculty Room in University Hall. And Irwin Shapiro, Director of the Harvard-Smithsonian Center for Astrophysics, announced the inauguration of a new annual lecture at the Observatory to be given in her honor and for which he would personally try to raise the endowment.

In the spring of 2000 Margaret Geller had brought the idea of a centenary celebration to the attention of the Center for Astrophysics administration, and the concept of a symposium was framed in relatively short order. Joining me on the Organizing Committee that selected the speakers were Andrea Dupree, Katharine Haramundanis, David Latham, A. G. Davis Philip and Virginia Trimble.

David Latham managed the web site and Dave Philip organized this volume for the press. The poster, a miniature of which is included here, was prepared by John Hamwey. Nancy Galluccio handled the financial side of the centenary celebration, Nayla Rathle and Sara Yorke assisted in planning for the logistics of the symposium, and Marion Tomasello in the Smithsonian Conference Office arranged the travel and housing for the speakers. Leslie Feldman, Betty Wickis, Lisa Oray Linarte and Valerie McCollom managed the pre-meeting registration procedures, and Jean Collins and Peggy Herlihy helped them staff the registration desk on the days of the meeting.

Irwin Shapiro obtained major financial support for the symposium and this volume from the Perkin Fund. The internal fund raising was spearheaded by Peggy Herlihy, with significant encouragement from Ramesh Narayan, Chairman of the Harvard Astronomy Department. Significant funding came from the Harvard Astronomy Department including a special allocation from the Dean's Office and from the University's Office of the Vice President of Administration. The Smithsonian Astrophysical Observatory supported the travel of the speakers, while the Ann Radcliffe Fund of Harvard College enabled selected undergraduates to participate.

Owen Gingerich

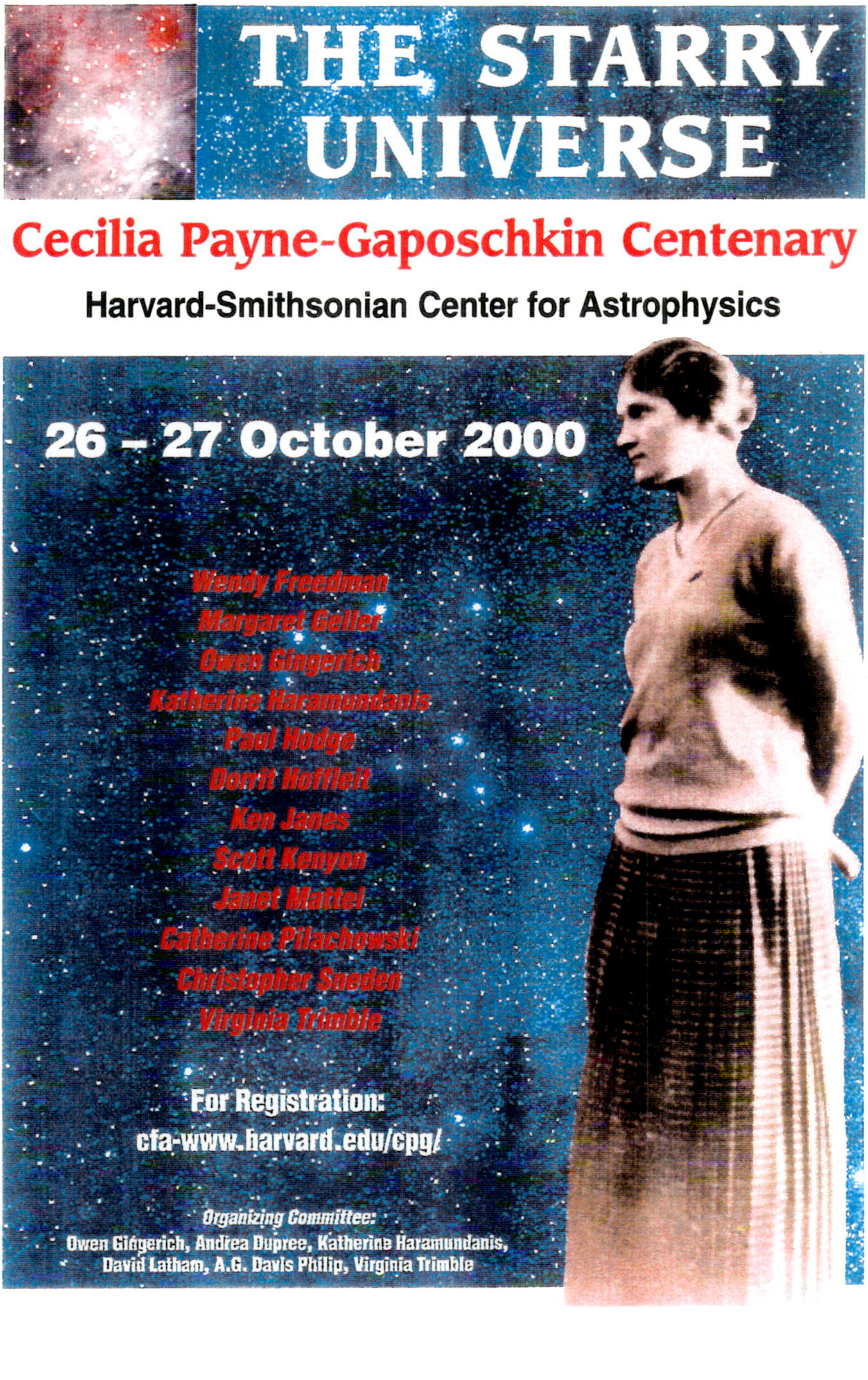
THE STARRY UNIVERSE
Cecilia Payne-Gaposchkin Centenary
Harvard-Smithsonian Center for Astrophysics
26 – 27 October 2000
Wendy Freedman
Margaret Geller
Owen Gingerich
Katherine Haramundanis
Paul Hodge
Dorrit Hoffleit
Ken Janes
Scott Kenyon
Janet Mattei
Catherine Pilachowski
Christopher Sneden
Virginia Trimble
For Registration:
cfa-www.harvard.edu/cpg/
Organizing Committee:
Owen Gingerich, Andrea Dupree, Katherine Haramundanis,
David Latham, A.G. Davis Philip, Virginia Trimble

LIST OF PARTICIPANTS

Lori Allen	CfA
June M. Asato	AAVSO
Bob Asato	AAVSO
Jason Aufdenberg	CfA
Eugene Avrett	CfA
Irving W. Bailit, MD	AAVSO
Pauline Barmby	CfA
John Bean	Middlesex School
Sara Beck	AAVSO
Ray Berg	AAVSO
David Bersier	CfA
Rob Bogdasarian	Middlesex School
Angie Boyce	Harvard College
David Boyd	CfA
Sylvia Boyd	
Ann Bragg	CfA
Daniel E. Brannen (Jr.)	AAVSO
Nancy Brickhouse	CfA
Warren R. Brown	CfA
Barbara Carter	CfA
John H. Collins	AAVSO
Lew Cook	AAVSO
Kesiah Cook	Harvard College
David Cowall	AAVSO
Tom Cragg	AAVSO
Mary Cragg	AAVSO
Mark Daranciang	AAVSO
Darryl Davis	Museum of Science
Kate Davis	AAVSO
Elizabeth Davis	Daughter of Donald Menzel
Peter Desloge	Middlesex School
William G. Dillon	AAVSO
Andrea Dupree	CfA
Mary Dussault	CfA
Richard Edgar	CfA
Martin Elvis	CfA
Ruth Esser	CfA
Nancy Evans	CfA
Giuseppina Fabbiano	CfA
Ted Farwell	Middlesex School
Michael Fisher	Harvard University Press
Elaine Fortin	CfA
John Fox	Harvard College
Wendy Freedman	Carnegie Observatories
Terry Gaetz	CfA
John Galloway	Endicott College
Thom Gandet	AAVSO

Margaret Geller	CfA
Tom Gilmore	Middlesex School
Miriam Gingerich	Cambridge
Owen Gingerich	CfA
Alyssa Goodman	CfA
Jonathan Grindlay	Harvard University
Sandy Grindlay	Harvard University
Andrew Hamill	Middlesex School
Kathy Haramundanis	Compaq, daughter of Cecilia
B. A. Harris	Harvard University
D. E. Harris	CfA
George Hawkins	AAVSO
Martha Hazen	CfA
Dudley Herschbach	Harvard University
Georgene Herschbach	Harvard University
Anna Hillier	Amateur Telescope Makers of Boston
Alan Hirshfeld	University of Massachusetts - Dartmouth
Paul Hodge	University of Washington
Dorrit Hoffleit	AAVSO & Yale University
Cap Hossfield	AAVSO
Coleman Hoyt	Middlesex School
Richard Hunt	Harvard University
Dave Hurdis	AAVSO
Gaitee Hussain	CfA
Kenneth Janes	Boston University
Hannah Jang-Condell	CfA
Christine Jones	CfA
Regina A. Jorgenson	Maria Mitchell Observatory
Dan Kaiser	AAVSO
Renate Kalkofen	CfA
Wolfgang Kalkofen	CfA
Vicky Kalogera	CfA
Sheila Kannappan	CfA
Sharon Kay	Massachusetts Institute of Technology
Patrick Kelly	Harvard College
Dan Kennedy	
Scott Kenyon	CfA
Steve Kilston	Ball Aerospace Technologies Corporation
Sungeun Kim	CfA
Kate Kirby	CfA
Jacqueline S. Kloss	Radcliffe
Rebecca A. Koopmann	Union College
Andrea Kurtz	Harvard College
Michael Kurtz	CfA
Robert Kurucz	CfA
Adair Lane	CfA
Kristine Larson	AAVSO
David W. Latham	CfA
Margaret E. Law	Harvard College

Rudi Paul Lindner	University of Michigan - Ann Arbor
Tiffany Llenos	
Alex Lobel	CfA
Rudy Loeser	CfA
Edward Los	AAVSO
Gilbert Lubcke	AAVSO
Andisheh Mahdavi	CfA
Kerri Malatesta	AAVSO
Bob Manske	AAVSO
Massimo Marengo	CfA
Eric Martin	CfA
Ursula Marvin	CfA
Janet Mattei	AAVSO
Alan Maxwell	CfA
Bruce McHenry	CfA
Haldun I. Menali	AAVSO
Gamze Menali	AAVSO
Florence Menzel	wife of Donald Menzel
Doug Mink	CfA
Jim Moran	CfA
Philip Morrison	Massachusetts Institute of Technology
Phyllis Morrison	Massachusetts Institute of Technology
Mario Motta, MD	AAVSO
George Mumford	Tufts University
Ruth Murray	Harvard College
Ramesh Narayan	CfA
Larry Nathanson, MD	AAVSO
Joy Nichols	CfA
Charley Noecker	Ball Aerospace Technologies Corporation
Michael Pahre	CfA
Barbara Palmer	CfA
Jay M. Pasachoff	Williams College
John Pazmino	AAVSO
Rebecca Pellock	AAVSO
Carolyn Collins Petersen	AAVSO
A. G. Davis Philip	ISO and Union College
Kristina Philip	ISO
Catherine A. Pilachowski	NOAO
Richard Porter, MD	AAVSO
Chuck Pullen	AAVSO
Ken Rines	CfA
Arthur Ritchie	AAVSO
Leif J. Robinson	Sky and Telescope
Joshua Roth	Sky and Telescope
Ronald R. Royer	AAVSO
Vera Rubin	Carnegie Institution
George Rybicki	CfA
Dimitar Sasselov	CfA
Sheila Sasselov	CfA

Sara Schechner	Harvard College
Irwin Shapiro	CfA
Julie Shaw	CfA
Mike Simonsen	AAVSO
Howard Smith	CfA
Kyle Smith	Middlesex School
Chris Sneden	University of Texas
Peter Sollins	CfA
Bo Stanek	Harvard College
Kris Stanek	CfA
Simon Steel	Harvard University
Vladimir Streinitski	Maria Mitchell Observatory
Gene Stubbs	
Angela Swafford	Knight Science Journalism Fellow - MIT
Andrew Szentgyorgyi	CfA
Patrick Thaddeus	CfA
Donna Thompson	CfA
Susan Tokarz	CfA
John Toone	AAVSO
Andrew Torres	Harvard College
Guillermo Torres	CfA
Jennifer Toy	Harvard College
Wes Traub	CfA
Virginia Trimble	University of Maryland, UC Irvine
David Turner	AAVSO
Arthur R. Upgren	Wesleyan University
Joan Upgren	Middletown
Alex Vanderburgh	Boston
Herbert Vetter	First Parish in Cambridge
Saeqa Dil Vrtilek	CfA
Elizabeth Waagen	AAVSO
Henry T. Wadzinski	Pracownia Kitlowskiego
William H. Waller	Tufts University
Martha Warren	AAVSO
Wayne Warren	AAVSO
Patricia Watwood	Artist for CPG portrait
Frederick R. West	AAVSO
Margaret West	AAVSO
Carol Weston	New England Aquarium
Fred Whipple	CfA
Winston S. Wilkerson	AAVSO
Lee Anne Willson	Iowa State University
Christine Wood	CfA
David E. Wright	AAVSO
James Wright	NSF
William F. Wyatt	CfA
Peter Young	CfA

Picture of the participants at "The Starry Universe". (Taken by Ronald Royer)

CONTRIBUTED PAPERS

Session Chairs

Thursday Afternoon	Andrea Dupree
Colloquium	David W. Latham
Banquet	Owen Gingerich
Friday Morning	Dimitar Sasselov
Friday Afternoon	Alyssa Goodman

"THE MOST BRILLIANT Ph.D. THESIS EVER WRITTEN IN ASTRONOMY"

Owen Gingerich

Harvard Smithsonian Center for Astrophysics, Cambridge, Massachusetts

ABSTRACT: Otto Struve's 1962 evaluation of Cecilia Payne's 1925 "Stellar Atmospheres" was based on her brilliant demonstration that the varied appearance of the stellar spectral types depended primarily on temperature differences whereas the chemical composition of the stars was essentially homogeneous. This article, by a former advisee, uses archival sources to describe the circumstances leading to her remarkable thesis.

I first arrived at Harvard College Observatory just over fifty years ago. I was a very green undergraduate, a summer assistant for Dr. Shapley – that's how the entire staff referred to him – and I had an office across from his in Building D. Mrs. G. – that's what everyone called Cecilia Payne-Gaposchkin – seemed a formidable, rather remote presence, "of imposing stature and stormy personality" as Jesse Greenstein characterized her.[1] I can't remember where she sat, but Sergei – that's how we all referred to her feisty husband – had his office on the stairs up to the 15" telescope, and the graduate students assumed that their two offices were widely separated to keep the stormy personalities from asserting themselves too conspicuously.

Those initial impressions seem curiously at odds with my later memories, for though Mrs. G. was always formidable in both intellect and stature, she soon struck me as gentle and kind. Our paths came together in my first semester in graduate school, when I enrolled in her "Introduction to Observational Astrophysics". She taught the course in Byerly Hall in the Radcliffe Yard, where all the astronomy classes open to undergraduates then met. This is where I really learned what "chain smoking" was. A pack of cigarettes and a single match could get her through the entire period. Basically hers was a course about stellar spectroscopy. Her teaching fellows were kept very busy producing packs of photographic spectra for us – of Eta Carina, T Corona Borealis, Zeta Puppis and Nova Herculis, of sunspots and the Lick moving spectra of the flash spectrum, from anywhere that suitable examples could be found. She produced for us extensive lists of lines, and she taught us how to use Charlotte Moore's multiplet tables. She carefully marched us through one fascinating spectrum after another. The individual stars were her friends, and so were the spectral lines. She probably knew most of them by memory.

Mrs. G. told me how she had been converted to astronomy. She had heard a lecture by Arthur Eddington on relativity, and had been so impressed that she went back to her room and copied it all out. "I know I got it right", she added slyly, "because he later published his lecture".

The Starry Universe: The Cecilia Payne-Gaposchkin Centenary
A. G. Davis Philip & Rebecca A. Koopmann, eds.
p. 3-16

Owen Gingerich
Harvard Smithsonian Center for Astrophysics
60 Garden St., Cambridge
Massachusetts 02138
GINGER@CFA.HARVARD.EDU

For the observational astrophysics class she borrowed a spectacular plate of a Zeeman spectrum from MIT, and it fell to me to analyze it for my class project. Years later, after she had died, I wanted to publish a sample of a Zeeman spectrum, but I had forgotten not only what atom it was, but the wavelength range as well. I sought help from Charlotte Moore Sitterly and others, but in vain. Finally one of Alex Dalgarno's colleagues in New York said, "I can't identify the atom, but the j values of the transitions are 4½ - 5½ and so on down". That was all I needed to identify it as vanadium, multiplet 22 in the multiplet tables, and I felt very smug about Mrs. G.'s legacy – there can't have been very many other students in the country trained to handle such a problem. She always loved puzzles like that.

My enthusiasm for this spectrum sealed my path through graduate school – clearly I was destined to be her thesis student, but some of the same circumstances that had eventually pulled her away from her first love, spectroscopy, also posed serious obstacles for me. Early in this century Harvard Observatory had gotten a terrific start mass-producing spectra for a large sky survey, but they fell behind in obtaining the higher dispersion spectra that she would have loved to analyze. In the 1950s Harvard was not a congenial place for budding spectroscopists, being unable to compete with the Western observatories. She begged and borrowed plates, but they were hand-me-downs, not assembled with challenging, cutting-edge problems in mind, and thus it was that eventually I turned to the computation of stellar atmospheres with Chuck Whitney using the latest in high-speed computing – a giant machine with 8000 words of memory – rather than working on curves of growth for Arcturus. Nevertheless, I kept in close touch with Mrs. G., learning her views on Italian art or paleolithic axes or mosaic woodworking or the earliest printed edition of Reynard the Fox, all topics that deeply interested her. Occasionally she reflected historically on her admiration for Eddington or how Dr. Shapley was the best conversationalist she had ever met. "Everyone adored him", she wrote in her autobiography. "Adelaide [Ames] and I called him 'the Dear Director', and soon he was affectionately known as 'the D. D'."[2]

"In spite of a vigorous scientific companionship, Dr. Shapley kept his distance. He never forgot, or let me forget, that he was the Director of the Observatory. I knew him for more than 50 years, and never once did he call me by my first name." Personally I don't know of anyone who called Dr. Shapley by his first name. From time to time he would set up tables in the Phillips Library in the form of a square, and the staff would gather round as he introduced visitors or expanded on the latest astronomical gossip. He called these sessions "Hollow Squares", but the graduate students referred to them as "Harlow Squares". There is a legend that when Cecilia Payne and Sergei Gaposchkin eloped in 1934, Dr. Shapley announced it at a Hollow Square and that Miss Cannon fainted dead away. In any event, I knew Mrs. G. for just over 30 years, more than half the time she spent at Harvard, but I don't recall if I ever called her by her first name. Nevertheless, in what follows I shall call her Cecilia simply to remind you that back in the 1920s she was a young graduate student in her 20s.

Cecilia Payne was born on May 10, 1900, in Wendover, England to a family with genealogical and intellectual connections to the intelligentsia of England. A box of letters she had inherited from her family included the autographs of Charles Darwin and the geologist Charles Lyell. Her educational trajectory took her to Newnham College in Cambridge, where she was allowed the unusual combination of botany, physics and chemistry. Though she was soon disillusioned by botany, her love of taxonomy found a natural niche in her ultimate pursuit of astronomy, where she eventually became a relentless classifier of variable stars.

But her interests turned more and more strongly to physics and then, inspired by Eddington's lecture on relativity, to astronomy. Among the courses she audited was one by the distinguished physicist Ernest Rutherford, and in her autobiography, "The Dyer's Hand", she records a wonderful anecdote:

"The stress was on observation. 'One thoroughgoing experiment,' Rutherford thundered in one of his lectures, 'is worth all the theories in the world...' Years later Eddington uttered the dictum that he would not believe an observation unless it was supported by a good theory. I was an astronomer by that time and knew him well. I told him I was shocked by his pronouncement. He smiled gently. 'I thought it would be good for Rutherford', he said."[3]

In May of 1922 the newly appointed director of Harvard College Observatory visited England and spoke at the Royal Astronomical Society. Cecilia, in the audience, was captivated, and after the lecture expressed to him her desire to study astronomy at Harvard Observatory. Nine months later she wrote:

> Dear Professor Shapley,
>
> I had the pleasure of seeing you at the meeting of the Royal Astronomical Society in May last, when I believe I said to you that I hoped to come and work at Harvard. I had decided some time previously that if it were possible I should wish to do research work under you. When, some months later, I consulted Professor Eddington, under whom I had been doing some work, as to what he would advise me to do if I were able to continue my studies at another University, he advised me most strongly to come to Harvard if I could.

This advice coincided completely with my own wishes and ambitions.

> I am extremely anxious to come if it is possible and am prepared to undertake anything that would enable me to work at Harvard. . . .[4]

Shapley encouraged her dream, but as Mrs. G. once confided to me years later, she thought he was quite startled when she actually showed up in the American Cambridge. But Shapley was not really taken by surprise, for there was a substantial correspondence preparing the way for her arrival. Included were fellowship recommendations from her mentors in England, and the common theme was her great enthusiasm. Eddington wrote, "I think her work shows much promise. . . . She . . . in addition possesses the valuable qualities of enthusiasm and energy in her work".[5] William Smart remarked, "It gives me great pleasure to testify to the abilities and enthusiasm of Miss Payne". Leslie Comrie observed, "Of her purely academic qualifications others are more competent to judge than I, but of her personal qualities of energy, perseverance and enthusiasm I can speak highly. . . . Should she be elected I feel certain that she will take her place worthily in the line of women astronomers that Harvard has nourished." G. F. C. Searle, University Lecturer in Experimental Physics, wrote, "She is a thoroughly earnest student and very keen on her work". And particularly touching is the recommendation from Frances Gray, head mistress of St. Paul's Girls' School: "It is not my practice to admit girls who have reached the age at which Cecilia Payne was admitted [age 17], but I was requested to make an exception in her case by the headmistress of the School she had previously attended, who assured me that she was a girl of very unusual promise."

> I soon saw for myself that this promise was likely to be fulfilled, and that Cecilia Payne had originality and ability far in advance of the originality and ability of the ordinary clever school-girl. I was very much struck, also, by her power of presenting a difficult subject in a lucid and attractive manner. I heard her give to the Science Club of the School a remarkably clear and interesting lecture on Aviation. It would not have been unworthy of an experienced teacher of Physics.

In responding to Shapley's encouragement, Cecilia wrote, "It is very good of you to say that I may work at Harvard under you, and I think it is a most inspiring prospect, which I shall do everything in my power to attain. . . .

I am extremely glad that you suggest the photographic study of variable stars as a possible line of work for me. I had hoped that you might suggest this subject, which has interested me perhaps more than any other branch, although I have done no original work upon it yet."[6]

In a letter that crossed hers Shapley again emphasized this kind of research: "As it is most likely that your work will lie along photometric lines, I should like to suggest that you familiarize yourself, if you have time before leaving England, with some of Miss Leavitt's and Professor Seares' work on photometric standards. . . ."[7]

By September of 1923 Cecilia had arrived on these shores. Soon she was busily working, not on photometry, but on spectra under Annie Cannon's tutelage. Before twelve months had passed she had won a class day prize (which took her by surprise) and she had gone to the British Association for the Advancement of Science meeting in Toronto, where Henry Norris Russell and Harvard's professor Frederick Saunders made sure that she met J. S. Plaskett, Ralph H. Fowler, Alfred Fowler and Georges Lemaître (who she described as "almost too good to be true – not shy, a sense of humour, enthusiasm and . . . surprising mental quality").[8] Her trip to Toronto included a highly memorable excursion to Niagara Falls, described to her fellow astronomy student Margaret Harwood:

> When we arrived I decided that solitude and silence were the only way to see the Falls, so I slipped away . . . and went off to stand by myself at the head of the Horseshoe Falls. . . . I don't know how long I was there, but I seemed to have been there always, when I turned around and found the only other person in the world whom I should have liked to be there, standing beside me. He also I suppose was after silence and solitude. I must say that was a good moment. . . .
>
> I went down to the foot of the Canadian Falls and was thrilled by the water coming down with such a roar; and I walked around everywhere and drank it all in; and didn't have to speak a word to a soul the whole time, (except for a probably inapposite but spontaneous comment to Eddington).[9]

Apparently Eddington went on after the Toronto meeting to lecture in Berkeley, and in November Cecilia got wind of the fact that he would speak at a neighborhood meeting at Yale as he headed back to England – "by accident – not from Shapley, who obviously isn't anxious to keep me up on this particular news. . . . perhaps Shapley wanted to spare me the pain of knowing it was going on at a meeting which my sex debars me from attending!"[10]

Meanwhile, in June of 1924, she had already taken a preliminary general examination for the Ph.D., set by Dr. Shapley and Prof. Saunders. The astronomical part of the exam survives in the Harvard Archives, together with a substantial part of her answers.[11] Among other things she was asked to write briefly on the Schaffierkassette, and briefly it was, for concerning the jiggle camera, she only wrote *"Schwarzschild method for photographic photometry. Difficulty of application to faint stars"* and on two-dimensional time she merely specified, *"Eddington (Math Theory of Relativity) An intellectual speculation of a daring kind – no apparent practical bearing of any sort"* but this was apparently enough to convince Shapley that she knew what the terms meant. She was equally dismissive of general relativity when asked about future astronomical research and the general theory of relativity. "*. . . the contacts between Relativity Theory and future astronomy do not appear to be very important*", she wrote. *"It does not . . . seem likely that much work can profitably be attempted with Relativity theory in mind. The theory may prove useful in elucidating observation but at the present stage can hardly guide it."*

The academic year 1924-25 was spent researching and writing her doctoral thesis. In her autobiography Mrs. G. reported that, concerning her degree, "One serious obstacle existed: there was no advanced degree in astronomy, and I should have to be accepted as a candidate by the Department of Physics. The redoubtable Chairman of that department was Theodore Lyman, and Shapley reported to me that he refused to accept a woman candidate."[12]

Be that as it may, the archival record seems far more routine. In September Lyman wrote,

> My Dear Shapley,
>
> Strictly speaking I have no right to sign Miss Payne's application for the degree of Doctor of Philosophy, but as I believe that some such action is generally taken in similar cases by the chairman of other Departments, I will gladly affix my signature and forward the paper to the Secretary of Radcliffe.[13]

In any event, organizing the degree for Cecilia *de facto* created the Astronomy Department.

Thus it was that at 2 p.m. Monday May 11 Cecilia had her final oral examination with Dr. Shapley and Prof. Saunders, who as an official member of the Radcliffe faculty had been delegated by Professor Lyman.[14] And by June, when she left for a vacation in England, she could write to Shapley from the steamship that "I left a complete MS of my thesis on the desk with the proof"[15].

Two months later bound copies of her monograph were available. Shapley reported to her in England that, "I have not looked through the book very critically, but my superficial examinations are sufficiently satisfying. I am particularly pleased with the binding. And of course the contents could be worse. I doubt if as good pieces of work have often before been put through in as short a time."[16]

No doubt Cecilia was used to Shapley's impish sarcasm by this time, and she responded from London, "I am glad you like the binding. I think it is admirable, and if inspection of the volume went only that far, I think it would create a good impression,"[17] but a week after his earlier letter Shapley felt modestly apologetic when he wrote, "To counteract some of the sarcastic disparagement by the Director, I am sending you a copy of the first page of a letter just received from Russell. He also writes in detail concerning some points – . . . None of them amount to a great deal – a perfectly astonishing revelation to me."[18]

Shapley's secretary, Arville Walker, typed out a copy of the first page of Russell's handwritten letter. He wrote as follows:

> My dear Shapley [they always addressed each other by their last names!]
>
> Please accept my hearty thanks for your letter, and for the presentation copy of Miss Payne's admirable book. It is the best doctoral thesis I ever read – with the exception, perhaps, of one on the orbits of eclipsing binaries! [an obvious reference to Shapley's own thesis written under Russell a decade earlier] – and ought to be strongly recommended, not only to observatory libraries, but to all students of the subject.
>
> As I read it over – I have eaten it up since I got it yesterday – I am especially impressed with the wide grasp of the subject, the clarity of the style, and the value of Miss Payne's own results. Many of these I knew before; of the others, the conclusion on p. 137 strikes me as especially pretty. To show that so much of the apparent discrepancy in the position of the maximum intensity is traceable to the relative abundance of atoms in higher and lower excitation states, and to have the computed pressures come out as they do after this correction, is very nice indeed.[19]

This was, in fact, the central brilliant result of Cecilia's thesis, so let me discontinue this review from the Harvard Archives and turn to the astrophysical background for the monograph itself and to an explanation of the significance of this major result.

Throughout the nineteenth century, the temperatures of stars was seen as a central unsolved astrophysical problem. Agnes Clerke devoted a chapter to the temperature of the Sun in her "A Popular History", and concluded, "Stefan's rule of a fourth-power relation, if not actually a law of nature, is a colourable imitation of one; and its employment has afforded a practical certainty that the Sun's temperature, so far as it is definable, neither exceeds 12,000° C., nor falls short of 6,500°

C".[20] The source of her numbers seems to be Charles Young, who in 1902 gave the same range from a value of the solar constant, that is, the solar energy falling each second on a square centimeter of the Earth's surface, and using Stefan's law.[21] A decade later we find C. G. Abbot using Stefan's, Wien's and Planck's laws to get a solar temperature between 5860° and 6260°, and he remarked that the spectral classes "indicate clearly a progression of temperature",[22] a statement that is hard to find in any of the Harvard sources, apparently because Director E. C. Pickering eschewed the mixing of theory with the empirical data of the Draper spectral classification project that had been carried out by Henrietta Leavitt and which had just passed into the hands of Annie J. Cannon.

By the early 1920s, especially following the pioneering work of Meghnad Saha in 1920 and '21, the first tables that correlated spectral type with temperature were published. There was much disparity between them, and in 1923 and 1924, when Cecilia began to work on this problem, it was indeed a hot problem in astrophysics. Her solution was so ingenious and satisfactory that it essentially turned the temperature problem into a non-problem, with the result that astronomers tended to forget the significance of this achievement and its consequences. What she did was to use the new quantum mechanical understanding of atomic structure, that is, the energy levels and how they are populated, to examine the temperature where a given atomic level would reach its maximum population. This is essentially a warfare between the ionization of the atoms, given by the Saha equation, and the population of the excited levels, given by the Boltzmann equation. For example, as the temperature of hydrogen gas is raised, more and electrons go into the n=2 level from which the Balmer spectrum so strong in the F and A stars arises, but as the temperature is raised still further, the hydrogen is ionized and there are fewer electrons left in the bound n=2 level, so in the hotter B stars the hydrogen lines decrease in intensity.

I once asked Mrs. G. why she had never mentioned the Boltzmann equation in her thesis. "Probably I had never heard of it", was her laconic reply. But indeed the equivalent equation, coming from Milne and Fowler, is essential to her calculations. She used these equations not only for hydrogen but for the other major elements with conspicuous lines in the visible region of the stellar spectra, and she established that despite the varied appearance of these spectra, their differences resulted essentially from the physical conditions and not from abundance variations. This chemical homogeneity of the starry universe is the essential point of her thesis, and the conclusion that Russell immediately zeroed in on. It is this result that Otto Struve highlighted when he called her book, "Stellar Atmospheres", "undoubtedly the most brilliant Ph.D. thesis ever written in astronomy".[23] It was a stunning insight into the uniformity of nature in the chemical construction of the universe.

But there is more. Following Part II, called "Theory of Thermal Ionization", there is Part III, "Additional Deductions from Ionization Theory", and in a short chapter entitled "The Relative Abundance of the Elements" there is a ticking time bomb. This is the extremely high abundance of hydrogen and helium that had come out under certain assumptions in the analysis.

I ask you now to do something extremely difficult, to forget your modern prejudices and your 20-20 hindsight, and to go as historians into the year 1925 as you try to understand how to interpret some extraordinarily strange results that fly in the face of common understanding. Miss Payne has made a definitely unproven assumption, that all the stronger lines at marginal appearance or disappearance as one moves along the spectral sequence correspond to the same number of atoms in that state of ionization. "This hypothesis is not as wild as we might suppose at first."[24] That is a direct quotation from Arthur Eddington's 1926 classic, "The Internal Constitution of Stars", a quotation that clearly delighted her, because Mrs. G. mentioned it to me more than once. But her assumption yields a high abundance for hydrogen and helium, an apparent anomaly with respect to the assumed homogeneity of the solar system. After all, when the Earth is taken as a whole, it must be predominately iron in order to account for its high mean density, and this is supported by the fact that meteorites are largely iron and by the appearance of the solar spectrum itself, which shows more lines of iron than any other element. The very important principle of uniformity of nature seems at stake here. As Cecilia herself argued in her thesis, "If . . . the Earth originated from the surface layers of the Sun, the percentage composition of the whole Earth should resemble the composition of the solar (and therefore of a typical stellar) atmosphere. . . . Considering the possibility of atomic segregation both in the Earth and in the star, it appears likely that the Earth's crust is representative of the stellar atmosphere."[25] So in her final table of abundances, she omitted hydrogen and helium, and it is this table that Eddington reproduced in his book.

How, then, to account for the anomalous abundance of these two light elements? Cecilia dismisses them in a single paragraph: "The outstanding discrepancies between the astrophysical and terrestrial abundances are displayed for hydrogen and helium. The enormous abundance derived for these elements in the stellar atmosphere is almost certainly not real. Probably the result may be considered, for hydrogen, as another aspect of its abnormal behavior, already alluded to; and helium, which has some features of astrophysical behavior in common with hydrogen possibly deviates for similar reasons."[26]

While Cecilia was deep in her thesis research, she took off time in the first days of 1925 to go to the AAS/AAAS meeting in Washington, the famous meeting where Hubble's results on the distance to M 31 were announced. "The D. D. had sent me off with the admonition to 'breeze up' to people", she wrote to her friend Margaret Harwood after her return, "and I thought Brown, of whom I am terrified, would be a good one to start on".[27]

> Brown encouraged me to join the party . . . and even offered me a cigarette, which I scandalized him by accepting. . . .
>
> After two days I decided to begin "breezing up" as nothing seemed likely to come my way otherwise. From thenceforward I did not listen to any papers. . . . I spent my time selecting victims, oscillating from the Physical to Chemical section in quest of them, leaving astronomy to the other astronomers. In this way I saw J. Q. Stewart, to great effect; also Prof. Compton of Princeton from whom I gleaned much that was of value. The strain of introducing myself to the latter gentleman nearly finished me – but it had to be done. I also saw Smythe of Princeton – an old dancing partner of Cambridge (Eng.) days, who used to be too proud to talk shop, but who now ministered to my pride by recognizing me, and coming up on purpose to discuss one of my papers with me. . . .
>
> On the last evening I was worn out, and also a count of my money revealed the fact that I had not enough to buy me a dinner. You can imagine my gratitude when the D. D. asked me to dine with him and Russell, not only for the honour done me, but for the actual food (I was dreadfully hungry). Adelaide and I always noticed that the D. D. does tactful things like that, apparently on pure inspiration. He had to leave early, and I believe that Russell and I were the only remaining members of the party. After Russell had been broadcast, he came and joined me, and we talked the whole evening – about (can you guess?) poetry and ancient Rome. I should not have thought he was the same man, and I feel quite differently about him – certainly I shall not be afraid personally of him any more. (His power in the astronomical world is another matter, and I shall fear that to my dying day, as the fate of such as I could be sealed by him with a word.)

Many years ago I asked Mrs. G. why she had pulled back from what is in retrospect the correct solution to the hydrogen abundances. "Probably Henry Norris Russell talked me out of it", was her reply. It was in the Princeton Archives that I found what I call "the smoking gun letter", written about ten days after she had met him in Washington:

> January 14, 1925
>
> My dear Miss Payne:
>
> Here, at last, are your notes on relative abundance which you were so good to send me some time ago....

> You have some very striking results which appear to me, in general, to be remarkably consistent. Several of the apparent discrepancies can be easily cleared up. [Here Russell discusses Mg, Mg+ and K in some detail.]
>
> There remains one very much more serious discrepancy, namely, that for hydrogen, helium and oxygen. Here I am convinced that there is something seriously wrong with the present theory. It is clearly impossible that hydrogen should be a million times more abundant than the metals, and I have no doubt that the number of hydrogen atoms in the two quantum state is enormously greater than is indicated by the theory of Fowler and Milne. Compton and I sent a little note to 'Nature' about metastable states, which may help to explain the difficulty. . . .
>
> Very sincerely yours,
>
> Henry Norris Russell[28]

In his recent biography of Henry Norris Russell, David DeVorkin writes, "She certainly did not convince Russell, and if Russell knew anything, he knew that her argument would not convince others. . . . He was right. Payne probably knew this."[29]

To put DeVorkin's conclusion in a little richer context, we must examine more fully what else was going with respect to the hydrogen abundance in those years. A younger Princeton colleague of Russell's, John Quincy Stewart, co-author of the distinguished Russell-Dugan-and-Stewart textbook, was working on line broadening mechanisms in stellar spectra, and he too found indications of high hydrogen abundance.[30] Russell likewise tried to dissuade Stewart from this conclusion. Meanwhile, in Germany Albrecht Unsöld was examining the hydrogen line wings, and his detailed analysis, published in 1928, pointed to a high hydrogen abundance.[31]

Fifty years later, in an oral history interview with Unsöld, I suggested that he must have been one of the first to recognize the high hydrogen abundance. "Oh, that was much later", he responded. "All the astrophysicists were quite confused: was it a matter of abundance, a matter of line broadening, or strange excitation conditions? So, the problem of the hydrogen abundance remained quite unclear."[32] In any event, Unsöld left me with the impression that he believed his work, like Payne's, related only to the atmosphere of the Sun, and until the papers of Bengt Strömgren in 1932 and 1933, which worked out interior models using a high hydrogen abundance, there was no convincing demonstration that stars as a whole were largely hydrogen. After all, hydrogen being the lightest element, could have segregated up into the atmosphere, as Cecilia herself had alluded to, and as Svein Rosseland actively argued.[33]

Meanwhile, however, Russell, perhaps after visiting his recent doctoral student, Donald Menzel, at Lick Observatory in January of 1929 and seeing his results on the solar flash spectrum, reconsidered his stance on the hydrogen abundance, and produced that year a monumental ApJ paper entitled, "On the Composition of the Sun's Atmosphere".[34] DeVorkin points out that Russell's approach was entirely different from Payne's. Whereas she has concerned herself primarily with astrophysical evidence, Russell now argued from the *physical* nature of the hydrogen atom, and only then marshaled support from a variety of astrophysical data. Consequently, as DeVorkin says, "Only at the end of his seventy-one-page ApJ paper did Russell give full credit to Payne's 1925 conclusions, saying nothing of his original rejection of them. Now he showed how his results agreed with hers, which was 'very gratifying' because she had used different methods on her giant stars. A giant star evidently had an outer atmosphere of nearly pure hydrogen, 'with hardly more than a smell of metallic vapors in it.'"[35]

In the end it was Russell who had the connections and the maturity to bring the many threads of evidence together, and the prestige to persuade the community of the validity of this result. It was thus Russell who Atkinson credited for the large hydrogen abundance in his pioneering 1931 proposal that nuclear fusion could account for the energy output of the Sun, though he specifically cited Cecilia's conclusion that the proportions of the different elements seem to be roughly constant throughout the galaxy.[36] There was still much work to be done, and as Unsöld wrote to me in 1980, "As to the ratio of hydrogen to metals we could only attempt to disentangle the relevant mechanisms one by one from 1927 to 1940", and he drew attention specifically to the work of Rupert Wildt on the H^- ion, which for the first time explained the lack of a strong Balmer jump in the ultraviolet spectrum of the Sun, something that otherwise seemed to be an anomaly if hydrogen was really the predominant element in the solar atmosphere.[37]

Cecilia Payne's thesis had clearly played a seminal role in the development of astrophysics. As the young Otto Struve, then just an instructor at Yerkes Observatory, wrote in a lengthy review of her monograph for the ApJ in 1926, "Miss Payne's book is full of useful suggestions for the practical worker. Nearly every page contains references to problems which are open to investigation by the spectroscopist."[38] Like Moses, Cecilia had made a truly memorable contribution. And like Moses, she had glimpsed the promised land, but hadn't quite got there. With what we know today, we could wish that it were otherwise, but I believe my account tallies with the historical actualities, where it is the person who persuades his colleagues of a new result who gets the credit. But it's interesting to notice that, when in 1934 Princeton's president inquired with Henry Norris Russell about a possible staff member to groom as his replacement, Russell wrote that the best candidate in America "alas, is a woman!, – not at present on our staff", an obvious reference to Cecilia Payne.[39]

I once owned two copies of Cecilia's monograph. The first I took to her to autograph. She simply inscribed her name, with none of the flourish that celebrities or my more flamboyant colleagues are wont to use. As a historian, I decided to keep the other copy, once owned by Otto Struve, the sometime director of Yerkes Observatory who had called her monograph "the most brilliant Ph.D. thesis ever written in astronomy". I gave the one Mrs. G. had inscribed so meekly to one of my own thesis students. But the modesty of her inscription always reminded me of a philosophical stance that she held and which I still believe is dead wrong. "It doesn't really matter how you give the credit for a scientific discovery", she said, "for if one person doesn't discover it, someone else will". I've always wondered to what extent her own experience with her thesis reinforced that opinion.

In contrast, I strongly hold that individuals in history can make a difference, whether it is in the political, artistic, or scientific arena. I am sure that for many of us, Mrs. G.'s scientific work as well as her mentorship did make a unique difference, and for that reason I am very pleased to take part in this "Starry Universe" symposium in her honor.

END NOTES

1. Jesse L. Greenstein, "An introduction to 'The Dyer's Hand'," in Katharine Haramundanis, ed., "Cecilia Payne-Gaposchkin, an Autobiography and Other Recollections", Cambridge, 1984, p. 8.

2. Cecilia Payne Gaposchkin, "The Dyer's Hand: an Autobiography" 1979, in Katharine Haramundanis, ed. *op. cit.*, p. 155.

3. *Ibid.*, p. 117.

4. Courtesy of Harvard University Archives, Observatory Director's papers (Shapley), UA V 630.22 Box 7, Cecilia H. Payne to Harlow Shapley, 26 February 1923. Unless otherwise specified, the letters are from this box.

5. *Ibid.* (1923 letters of recommendation).

6. Payne to Shapley, 5 April 1923.

7. Shapley to Payne, 16 April 1923.

8. Payne to Shapley, 10 August 1924.

9. Payne to Margaret Harwood, Toronto, 16 August 1924. Harwood Archive, Schlesinger Library, Radcliffe Institute, Harvard University. I am indebted to David DeVorkin for pointing out these letters.

10. Payne to Harwood, 22 November 1924. By permission of Harwood Archive, Schlesinger Library, Radcliffe Institute, Harvard University.

11. A copy of two of the three parts of the Ph.D. preliminary exam is found among Shapley's personal papers in HUG 4773.10.19c. Her answers to Shapley's part of the exam are found among the director's papers in UAV 630.22 (1920-29) Box 7 under "Gaposchkin". I am indebted to Robert Hall for this information.

12. Cecilia Payne-Gaposchkin. *op. cit.*, (note 2), p. 157.

13. Theodore Lyman to Shapley, 20 September 1924 (Box 12).

14. Shapley to Lyman, 6 May 1925 (Box 12).

15. Payne to Shapley, 10 June 1925.

16. Shapley to Dr. Cecilia Payne, 6 August 1925.

17. Payne to Shapley, 20 August 1925.

18. Shapley to Payne, 15 August 1925.

19. Henry Norris Russell to Shapley, 11 August 1925.

20. Agnes M. Clerke, "A Popular History of Astronomy during the Nineteenth Century", London, 1902, p. 226

21. Charles A. Young, "Manual of Astronomy", Boston, 1902, p. 252. In the revised edition of his "A Text-Book of General Astronomy" Boston, 1900 he was still far from these numbers.

22. Charles Greeley Abbot, "The Sun", 1911, p. 410.

23. Otto Struve and Velta Zebergs, "Astronomy of the 20th Century", New York, 1962, p. 220.

24. Arthur S. Eddington, "The Internal Constitution of the Stars", Cambridge, 1926, p. 369.

25. Cecilia H. Payne, "Stellar Atmospheres", Harvard Observatory Monograph No. 1, Cambridge, 1925, p. 185.

26. *Ibid.*, p. 188.

27. Payne to Margaret Harwood, 9 January, 1925 Schlesinger Library, Radcliffe Institute, Harvard University.

28. Russell to Payne, 14 January 1925, Henry Norris Russell Papers, Series I: Outgoing Correspondence, Box 9, Folder 37; published with permission of the Manuscript Division, Department of Rare Books and Special Collections, Princeton University Library.

29. David DeVorkin, "Henry Norris Russell, Dean of American Astronomers", Princeton, 2000, p. 366.

30. John Quincy Stewart, "Popular Astronomy", 36 1928, 346. I am indebted to David DeVorkin for this reference; he remarks that no expansion of this abstract has been found.

31. Albrecht Unsöld, "Über die Struktur der Fraunhoferschen Linien und die quantitative Spektralanalyse der Sonnenatmosphäre", "Zeit. f. Physik", 46 1928, 765.

32. Oral history interview with Albrecht Unsöld by Owen Gingerich, 6 June 1978, American Institute of Physics, Niels Bohr Library.

33. Svein Rosseland, "On the Distribution of Hydrogen in a Star", MN 85, 1925, 541-46.

34. Henry Norris Russell, ApJ 70, 1929, 11-82.

35. DeVorkin, *op. cit.* note 29, p. 216.

36. Robert d'Escourt Atkinson, ApJ 73, 1931, 292.

37. Albrecht Unsöld to Owen Gingerich, 2 September 1980, personal collection.

38. Otto Struve, ApJ 64 1926, 204.

39. DeVorkin, *op. cit.*, note 29, p. 341.

CECILIA AND HER WORLD

Katherine Gaposchkin Haramundanis

Compaq Computer Corporation, Westford, Massachusetts

ABSTRACT: This collection of photographs, with a few additions, was shown at the CPG Centenary held 26-27 Oct 2000 in the Phillips Auditorium at the Center for Astrophysics, Harvard University, Cambridge, Massachusetts. Presented in chronological order, they illustrate a few facets of Cecilia Payne-Gaposchkin's very multi-faceted life.

1. EARLY YEARS

Cecilia was born on a spring day in May, when fruit trees and flowers were blooming in the bucolic town of Wendover, England. Wendover, Buckinghamshire, a moderate commute by train from London, had become the home of Emma and Edward Payne after their marriage in 1899. They named their house Holywell Lodge after the home of a patron, the Comtess de Noailles, who was especially fond of Emma. Emma, 33, a beautiful artist, had helped run the household of her grandfather J. J. Garth Wilkinson after the death of his wife, and made a major adjustment in her life with her marriage, as did her spouse Edward. Edward Payne, at 59, had spent twenty years at Oxford University, was an acclaimed historian (Payne 1892, 1899) and separately a barrister, and by family accounts was simply smitten by Emma, with the result that he proposed and they married. They were presented with their first child, Cecilia Helena, on 10 May 1900 [Fig. 1]. This is the earliest photo of Cecilia, taken at Wendover at about the age of one, in very elaborate Edwardian dress. Strikingly, the baby wears neither shoes nor socks.

In addition to his knowledge of history, Edward Payne was a well-known musicologist, contributing several articles to Grove's Dictionary of Music and Musicians, particularly on stringed instruments. Payne had traveled extensively in Italy, including Cremona, and had a collection of stringed instruments. Arthur F. Hill was one of his friends. Mr. Hill, author of a definitive biography of Stradivari (Hill et al. 1902) was a prominent British musicologist and violin maker. This was the milieu in which Cecilia spent her earliest years. An early photo [Fig. 2] captures a moment when the two families were together, with both Cecilia and her brother, two years younger than she. It shows Cecilia Payne, with her father, her

The Starry Universe: The Cecilia Payne-Gaposchkin Centenary
A. G. Davis Philip & Rebecca A. Koopmann, eds.
p. 17-66

Katherine Gaposchkin Haramundanis
P. O. Box 1365
Westford, Massachusetts
KATHY.HARAMUNDANIS@COMPAQ.COM

mother and Mr. Hill - at Wendover, about 1902. This appears to have been taken in the music room of their home Holywell Lodge, as three stops of their chamber organ are visible at the left. From left to right: Cecilia (on lap), Mrs. Hill(?), Emma Payne (mother), Humfry (held by Emma), Mr. Hill (seated), Edward Payne (father).

On Boxing Day (December 26) 1904, Edward Payne died suddenly, probably of a stroke or heart attack (Catch 1990, Payne 1892, 1899). He was found drowned in a canal in Wendover along the route for his morning walk. His third child Leonora, born that year, was yet a tiny baby. The inquest held the following day returned an open verdict, with no indication of foul play. Most likely he died of a heart attack. Earlier that year the two older siblings were captured in a studio photograph, in typical Edwardian dress [Fig. 3]. Their mother labeled this studio photograph as "Mother Cecilia 4 years old and Uncle Humfry nearly 2 years old". The photo was possibly intended for her grandson Edward, and delivered when Emma visited the United States in 1936.

The three children and Emma remained in Wendover for another eight years, participating in the activities appropriate for their class and upbringing. As was usual at that time, theatricals, tableaus, recitations and other forms of public performance were frequently part of their education. The Payne children were all fascinated by the theater, and acted in many plays as children and later. Their costuming and poses were sometimes captured in photos [Fig. 4]. From the costumes this looks like a medieval play but this could be "Ali Baba and the Forty Thieves" which Cecilia mentions in her autobiography.

Cecilia was already an eager student at the first of the schools she remembered so well, in the building just across the street from where they lived. This charming picture (Fig. 5) shows her looking over her mother's shoulder as they examine a periodical called "Animal World".

The house where the Paynes lived [Fig. 6] still stands today, though greatly encroached upon by new housing. It no longer stands among fields but in the midst of houses, though parts of Wendover remain charmly rural. The Paynes are both buried in the Parish Church of St. Mary the Virgin in Wendover, under a great stone book. The house remains a private residence and is now surrounded by a high fence. This photo shows it as it was when the Paynes lived there.

Cecilia had fond and very specific memories of her time at this school [Fig. 7], which still remains, across the street from their erstwhile home, though no longer as a school. It is currently a municipal building of the town of Wendover.

Cecilia had an extended family comprised primarily of aunts, among them two talented sisters who lived in Cambridge, England. Dorothea Pertz [Fig. 8], "dearest Auntie Dora" to the three Payne siblings, was the daughter of the important German historian and member of parliament, George Heinrich Pertz, great grandfather of Cecilia, and his second wife, Leonora Horner. Leonora's father, Leonard Horner, was an important Scottish geologist, and president of the Geological Society of London. Emma (Pertz) Payne was Dora's step grand-niece, and the family made an annual visit to Dora and her sister Annie, an artist, in the university town of Cambridge. Dora had been an early student at Newnham, and, trained as a botanist, she assisted in the laboratory of Sir Francis Darwin, a plant physiologist. He was the son of Charles Darwin, author of "Origin of Species and Descent of Man". Her example inspired Cecilia to consider that becoming a scientist was not impossible for a young woman.

In 1912 the family moved from Wendover to London so that all the children would have the advantages of better schools. This photo captures the three siblings in a tranquil pose [Fig. 9]. For Cecilia and Leonora, this was a severe uprooting, long remembered by them both, probably also for Humfry. A tragic consequence of this event was the death of their beloved dog, who could not be brought to London, and who pined away not long after their departure. But for Cecilia this time was the opening of new doors through which she was eager to fly. In this picture, Cecilia points to a place in the book with a finger of her left hand; perhaps this was done to indicate that she was naturally left-handed.

Humfry would fulfill his promise by becoming the Director of the British School of Archeology in Athens, Greece, while Leonora would make her mark as an artist with a specialty in architectural drawings, contributing many illustrations to books and newspapers.

Cecilia won several school prizes at St. Mary's College, which she attended at Lancaster Gate. One was the Works of Spenser, a second the poems of Coleridge, and one, which she describes in her autobiography as a treatise on fungi, was British Fungi, awarded for an Essay in Form IVa, July 1914. The prize in this picture [Fig. 10] from 1913, also includes her bookplate which is taken from a drawing her father used to illustrate his article on Stringed Instruments in Grove's Dictionary.

2. NEWHAM COLLEGE, CAMBRIDGE

Cecilia's educational prowess won her the support of several teachers, and eventually she was able to achieve her dream of being admitted to Newnham on a substantial scholarship. She worked sometimes to the point of exhaustion to achieve this, but achieve it she did. This studio portrait [Fig. 11] shows her at about the time when she entered Newnham in the fall of 1919.

Newnham College remains today much as it was at the time when Cecilia was there, although of course over so many years many changes have occurred. However, some of the college rooms remain much as they were then, and the grounds and buildings [Fig. 12] retain much of their flavor. There is something to be said for this remarkable school where at least two extraordinary women scientists have studied, Cecilia and Rosalind Franklin (1920 - 1958). Franklin, a molecular biologist and expert x-ray crystallographer, made X-ray photos that were the key to unraveling DNA. She should have been awarded the Nobel prize with Watson, Crick, and Wilkins, just as C. Payne should have been credited in the 1920s for her fundamental discoveries about the abundances of the elements.

Cecilia describes in her autobiography how she discovered the small visual telescope on the grounds of Newnham, and how she cleared it out and opened the observatory for public nights to observe the planets and stars, including variable stars. The observatory [Fig. 13] was recently refurbished by the Royal Engineers, and still stands on the Newnham grounds.

Cecilia remembered with feeling the lectures she heard from Rutherford and Bohr, [Fig. 14], who were developing atomic theory to which she would put immediate use in her astrophysical work. Rutherford was one of Cecilia's professors and Bohr a visiting lecturer. Lady Jeffreys quotes Rutherford as saying in 1925 about Cecilia: "My daughter's friend is doing well in astrophysics, an understatement in light of "Stellar Atmospheres" (Jeffreys 1985).

E. A. Milne [Fig. 15], a Cambridge astronomer, was a mentor and long time friend of Cecilia. Near contemporaries, they met at Cambridge and maintained their friendship until his untimely death in 1950.

After four heady years at Newnham, and inspired anew by lectures of Arthur Eddington and Harlow Shapley, Cecilia struck out for the New World to plunge into the great library of astronomical photographs at the Harvard College Observatory. She recounts the hilarity with which her dorm-mates observed her English style of clothing, and a photo taken at the time [Fig. 16] carries her own notation "C.H.P. in local costume".

The American Association of Variable Star Observers had space and held meetings at the Observatory during this period. One their meetings of 1923 was captured in this group photo [Fig. 17]. From left: Front row: Miss Cushman (1), M. F. Jordan (4), Miss Gushee (8), R. O. Suter (12) , Miss Woods (16), H. W. Culp (21), Miss Clough (22), D. B. Prentice (27), Miss Young (32), Harlow Shapley (35), Annie J. Cannon (39), C. C. Godfrey (43), Mrs. G. C. Waldo (46), Miss Howarth (50), P. R. Allen (52), Mrs. Holmes (54). Second row: Miss Hodgdon (2), Mrs. C. W. Elmer (in hat with feather) (5), Mrs.

M. F. Jordan (9), Miss Farnsworth (13), E. B. Jones (17), Miss Gill (23), Mrs. D. B. Pickering (28), Mrs. W. T. Olcott (33), Mrs. E. F. Sawyer (36) (last four in hats). Back row: R. N. Mayall (3), W. W. Bickell (6), M. J. Jordan (7), Mr. Burley (11), W. D. MacPherson (10), G. C. Waldo (14), Leon Campbell (15), J. E. G. Yalden (18), D. B. Pickering (19), Luis Rodés (20), C. T. Whitehorn (25), E. F. Sawyer (26), W. J. Luyten (29), M. Cilleyy (30), W. T. Olcott (34), C. Y. McAteer (37), Miss Ames (38), C. Payne (40), L. J. Boss (41), W. J. Reardon (42), Willam Henry (45), Miss Walker (44), Miss Murdock (47), J. P. Fox (48), S. B. Cummings (49), W. J. Fisher (51), W. H. Cudwirth (53), J. H. Metcalf (55). (Photo and identifications courtesy of AAVSO, Michael Saladyga, archivist; numbers in parentheses are identifications from Popular Astronomy, No. 310.)

CP was also present at the AAVSO meeting October 11, 1924, the following year at HCO [Fig. 18].

25 31 33 38 44 53 58

3 6 9 13 14 17 21 23 27 32 34 39 43 45 48 49 52 54 57 59 60 64

1 4 7 10 12 15 18 22 26 30 37 40 47 51 62 65

20 24

2 5 8 11 16 19 28 29 35 36 41 42 46 50 56 61 63

(1) King, (2) Mrs. Sawyer, (3)?, (4) Miss Hodgdon, (5) Campbell, (6) Brigham, (7) Mrs. M. F. Jordan, (8) Mrs. Elmer, (9) Butler, (10) Miss Farnsworth, (11) Tarwater, (12) Godfrey, (13) Cudworth, (14) Sawyer, (15) Miss Godfrey, (16) Cilley, (17) Miss Estabrook, (18) Miss Estabrook, (19) Le Maitre, (20) Miss Woods, (21) Miss Gill, (22) Suter, (23) Miss Cushman, (24) Prentice, (25) Henry, (26) Miss Partridge, (27) Miss Swartz, (28) Mrs. Morris, (29) Whittier, (30) Kearons, (31) Shapley, (32) Miss Maury, (33) Olcott, (34) Mrs. Farnham, (35) Yalden, (36) Mrs. Holmes, (37) Mrs. Kearons, (38) Miss Harwood, (39) Miss Leland, (40) M. J. Jordan, (41) Miss Young, (42) Miss Cannon, (43) Whitehorn, (44) Allen, (45) Miss Payne, (46) Brown, (47) Potter, (48) Mrs. Jewett, (49) Jewett, (50) Mowry, (51) Miss Potter, (52) Boss, (53) M. F. Jordan, (54) Mrs. Boss, (55) Elmer, (56) Fisher, (57) Stetson, (58) Murray, (59) Mayall, (60) Reardon, (61) Jones, (62) Miss Clough, (63) Fox, (64) Farnham, (65) Mrs. Rogers. (Photo and identifications courtesy of AAVSO, Michael Saladyga, archivist.)

In 1924, though not yet a graduate, Cecilia attended the meeting of the British Association for the Advancement of Science in Toronto, Canada. There she was able to meet with many prominent astronomers, including F. A. Saunders, J. S. Plaskett, A. Fowler, G. Lemaitre, E. A. Milne, Robert G. Aitken and A. Eddington. Some of these scientists are seen in a group photo from the event [Fig. 19]. (Front row, from left), -, G. Van Biesbroeck, Winifred Eddington, -, -, Joel Metcalf, -; (second row),

-, -, -, -, -, Margaret Harwood (?), A. Eddington, C. Payne, - ; (back row): -, H. N. Russell, -, Le Maitre, -, (? - blurred). Some identifications courtesy of Michael Saladyga, AAVSO archivist.

While in Toronto, Cecilia was able to visit the grounds of the University in front of Simcoe Hall [Fig. 20] as well as the botanical gardens. I am indebted to Dr. Nancy Evans, Center for Astrophysics, for the identification of this location.

At Harvard College Observatory, the ladies who were the backbone of the astronomical work are featured in this group photo [Fig. 21]. (Front row from left): Agnes M. Hoovens, Mary B. Howe, Harvia H. Wilson, Margaret Walton Mayall, Antonia C. Maury; (middle row): Lillian L. Hodgdon, Annie J. Cannon, Evelyn F. Leland, Ida E. Woods, Mabel A. Gill, Florence Cushman; (back row): Margaret Harwood, Cecilia Payne, Arville D. Walker, Edith F. Gill. (Courtesy of Harvard College Observatory.)

Cecilia continued to have an interest in botany, and visited the Botanical Gardens just across Garden Street from the Observatory. For this photo [Fig. 22], her caption is "In hat with green ribbon in front of the Botanical Garden".

Taken in about 1925, this photo [Fig. 23] shows Cecilia in a slightly formal pose and stylish clothing. She is probably holding a cigarette, to which habit she had been introduced by her brother.

Cecilia's time as a graduate student was completed with the publication of the first monograph of the Harvard Observatory [Fig. 24], which was her Ph. D. thesis. She had published several papers before this, culminating in this extraordinary work that gave definitive results for the abundances of the elements in stars based on her observations and her use of atomic theory, then in its infancy.

As she recounts in her autobiography, for a meeting of astronomers at the Observatory, the staff put on a performance of the "Observatory Pinafore" which spoofed the astronomers and their zealous pursuit of science [Fig. 25]. In the Director's Residence, Harvard College Observatory: From left: ?, Bart Bok, ?, Prof. Ransom from Tufts, CP (holding the "vexatious prisms"), Peter Millman, Henrietta Swope, Frank Bowie, Helen Sawyer, Arthur Sayer, ?, Leon Campbell.

At the end of the 1920's, Frances Wright came to the Observatory to work in astronomy, and she and Cecilia became good friends. They often went on outings together. In this photo [Fig. 26] they are seen on a walking excursion.

3. TRAVELS

Having become good friends, and united in a common sense of adventure, Cecilia and Frances shared an apartment and decided on an auto trip to the western observatories. They drove in a car [Fig. 27] that surely had a manual shift, explored their way on unpaved roads west of the Mississippi, camped, and met adventure as it happened. Their auto was a convertible, with wire wheels and wide running boards used for extra baggage.

They took a side trip by mule to the bottom of the Grand Canyon. Fig. 28 shows Cecilia at the top of Bright Angel Trail with a trail guide in full western dress. I am indebted to a member of the audience at the CPG centenary who identified this location. Frances is captured on her mule half-way down the same trail [Fig. 29].

Cecilia returned home to Britain whenever possible, and Fig. 30 shows her with her mother and sister on an outing there. They are all very fashionably dressed, and Leonora holds a sophisticated camera. From the left: Emma, Leonora and Cecilia.

In 1933 Cecilia decided to travel to the northern observatories of Europe and to visit Boris Gerasimovich, Director of the Pulkova Observatory, near Leningrad in the Soviet Union. Although many tried to dissuade her from making this arduous journey, she persevered and eventually reached her destination. She took this photo [Fig. 31] of Gerasimovich as they surveyed the grounds of the observatory from one of the higher buildings. Cecilia also photographed several buildings and structures of the observatory [Figure 32] on the Pulkova Heights. These buildings were largely destroyed in the war that reached this area in 1941.

At the end of her trip to the northern observatories, Cecilia attended the meeting of the Astronomisches Gesellschaft in Göttingen, Germany. The group photo of this event [Fig. 33] shows Cecilia standing on the landing. It was at this meeting that Cecilia met Sergei Gaposchkin, then a refugee for the second time. He had left Russia by accident during their civil war, pursued his education in Bulgaria and Germany, studied at the Kaiser Wilhelm University in Berlin, where Albert Einstein ran the Institute for Physics, and completed his doctoral dissertation in 1932. With the fall of the Weimar Republic and the accession to power of Adolf Hitler, Sergei lost his position at the university observatory in Neubabelsberg. Interviewed by a Nazi official, he believed that he was scheduled to enter the Sonneberg concentration camp. The Soviets were suspicious of a man who had lived in Germany for a decade and refused him permission to enter the Soviet Union. From a colleague he heard about the meeting of the Astronomisches Gesellschaft and bicycled to Göttingen in hopes of finding employment outside of Germany. He did not really count on moving to America, but Cecilia brought his case to the attention of Harlow Shapley who offered him a position as a research assistant. Leaving Germany on a stateless passport, with a quickly obtained visa from the American Consul in Berlin, he arrived in Boston on the Georgic in October 1933.

The Astronomische Gesellschaft was held at Universitats-Sternwarte, Göttingen (USG), August 1933 (Vierteljahresschrift 1933). I am greatly indebted to Willi Deinzer and Axel Wittmann at USG for their identifications of the participants. Files and pictures were prepared by Axel Wittmann (December 2000).

a) List of persons by number (see labeled group photograph):

001 Richard Prager (certain), 002 Paul Guthnick (certain), 003 Friedrich Wilhelm Ludendorff (certain), 004 August Kopff (certain), 005 Herzog Ernst v. Sachsen-Altenburg (certain), 006 Otto Knopf (certain), 007 Elis Strömgren (certain), 008 Knut Lundmark (certain), 009 Josef Hopmann (certain), 010 Johannes Wempe, 011 Alexander Wilkens (certain), 012 Albrecht Kahrstedt (certain), 013 Axel Nielsen (certain), 014 Johannes Larink (certain), 015 Heinrich Siedentopf (certain), 016 Carl Adam Schalén (certain), 017 Otto Schlier (certain), 018 Josef Weber, 019 Karl Boda, 020 Hans Kienle (certain), 021 Thomas Gengler (certain), 022 Otto Heckmann (certain), 023 Mrs. Heckmann (certain), 024 Arnold Kohlschütter (certain), 025 Ludwig Biermann (certain), 026 Harald von Klüber (certain), 027 Bruno Thüring, 028 Cuno Hoffmeister (certain), 029 Erich Schönberg (certain), 030 Karl Stumpff (certain), 031 Rolf Müller, 032 Martin Schwarzschild (certain), 033 Rupert Wildt (certain), 034 Cecilia Payne (certain), 035 Richard Schorr (certain), 036 Bengt Strömgren, 037 Paul Hügeler (?), 038 Giuseppe Armellini (certain), 039 Hermann Brück (certain), 040 Karl Pilowski (certain), 041 Margarete Güssow (certain), 042 Joachim Stobbe (certain), 043 Kurt Ruthe (certain), 044 Max Mündler (certain), 045 Arthur Eddington (certain), 046 Friedrich Gondolatsch (certain), 047 B. Sticker, 048 M. Lindow (certain), 049 Julie Vinter Hansen, 050 Hans Strassl, 051 Dorothea Roberts-Klumpke (certain), 052 Willy Hartner (certain), 053 Wilhelm Becker (certain), 054 Hermann Kobold (?, unregistered participant ?), 055 J. Hellerich (?), 056 Andreas Gottfried Galle (certain), 057 Ernst Anding (certain), 058 Ludwig Terkßn, 059 Sigmund Mauderli (certain), 060 Sergei Gaposchkin (certain), 061 Julius Dick (certain), 062 Erwin Roux (?), 063 Hermann Strebel (certain), 064 Walter Villiger (certain), 065 ?, 066 ?, 067 ?, 068 Johannes Peters (certain), 069 Johann Behrens (certain), 070 ?, 071 ?, 072 ?, 073 ?, 074 ?, 075 Johann Wilhelm Stein, 076 ?, 077 Carl Vick (certain), 078 Josef Weber, 079 Werner Schaub(?) (or Oswald Wachtl (?)), 080 ?, 081 Kurt Walter(?) (or Bertil Lindblad(?) or Karl Reinmuth(?)), 082 ?, 083 ?, 084 ?, 085 Gabriella Armellini-Conti, 086 ?, 087 ?, 088 ?, 089 Paul Hügeler (certain), 090 ?, 091 ?, 092 ?, 093 ?, 094 Arthur König (?), 095 ?,

096 ?, 097 ?, 098 ?, 099 ?, 100 Rolf Müller, 101 Heinrich Vogt, 102 ?, 103 Fritz Goos, 104 ?, 105 Max Beyer, 106 Martin Ebell (certain), 107 ?, 108 Paul Labitzke, 109 ?, 110 ?, 111 Bruno Meyermann (certain), 112 ?

b) List of participants in alphabetical order (merged from the official list of participating AG members and the signatures - including those of non-members, wives of participants, etc. - in the guest book). Entry 'ident. not found' means that a portrait was available but the person could not be found or identified in the group photograph. Note that some of the signatures were difficult to read, so the spelling of some of the names is uncertain. Dates of birth are given to assist in identifications; [p.x] means signed the guestbook on page x; (y) is the identification number in the photo.

Ruben Andersen, Aarhus, DK, b. 1892, [p. 3]; Ernst Anding, Gotha, b. 1860, (57); Gustav Angenheister, Göttingen [p. 5]; J. Armeanca, Cluj, RO, [p. 6]; Giuseppe Armellini, Roma, IT, b. 1887, [p. 1], (38); Gabriella Armellini-Conti, Roma, [p. 1], (85 ?); Siegfried Baumerly (?), Kiel, [p. 5]; A. Frh. von Beaulieu Marconnay, Berlin-Wilmersdorf; Friedrich Becker, Bonn, b. 12.06.1900, [p. 1], (87 ?); Maria Becker, [p. 1]; Wilhelm Becker, Potsdam, b. 03.07.1907, [p. 6], (53); Hanny Behrens, Stâde, [p. 4]; Johann Gerhard Behrens, Stade, b. 1899, [p. 4], (69); Walter E. Bernheimer, Wien, AU, b. 1892, (ident. not found); Anneliese Beyer, Hamburg, [p. 4]; Max Beyer, Hamburg, b. 22.10.1894, [p. 3], (105); Ludwig Biermann, Göttingen, b. 13.03.1907, [p. 5], (25); Gretl Birnheimer-Stayer, Wien, [p. 6]; Karl Boda, Frankfurt/Main, b. 1889, [p. 4], (19); Kurt Felix Bottlinger, Berlin-Babelsberg, b. 1888, [p. 1], (ident. not found); Hermann Alexander Brück (Potsdam), b. 15.08.1905, [p. 6], (39); A. von Brunn, Potsdam; Ludwig Burg / Bürg (?), Jena, [p. 1]; H. Dahmaschiscke (?), [p. 5]; Julius Hermann Gustav Dick, Berlin-Babelsberg, b. 1891, (61); August Dose, Kiel, b. 12.01.1902, [p. 6], (ident. not found); Martin Ebell, Kiel, b. 1871, [p. 1], (106); Arthur Stanley Eddington, Cambridge, UK, b. 1882, (45); Herbert Eggen, [p. 7]; Michael Esch S.J., Valkenburg-Limburg, NL, b. 1869, (ident. not found); H. Friedrichs, Göttingen, [p. 7]; Lore Galbas, Essen, [p. 2]; P.A. Galbas, Essen, [p. 2]; Andreas Wilhelm Gottfried Galle, Potsdam, b. 1858, [p. 2], (56); G. Thomas Gengler, Göttingen, b. 1901, [p. 7], (21); Luise (?) Giese, Göttingen, [p. 7]; H. Goldschmidt, Göttingen, [p. 4]; M. Goldschmidt, Göttingen, [p. 4]; Friedrich Gondolatsch, Berlin-Dahlem, b. 03.06.1904, (46); Fritz Goos, Hamburg, b. 1883, [p. 2], (103 ?); H. Gräfe, [p. 6]; H. Gresch (?), Göttingen, [p. 3]; S. Günther, Göttingen, [p. 3]; Margarete Güssow, Berlin-Babelsberg, b. 1896, [p. 2], (41); Paul Guthnick, Berlin-Babelsberg, b. 1879, (2); Gerry Hart(w)ig, Nehlstädt/Thür., [p. 3]; Willy Hartner, Bad Homburg, b. 22.01.1905, (52); R. Hase, Hannover, [p. 4]; Otto Heckmann, Göttingen, b. 23.06.1901, [p. 7], (22); J. Hellerich, Hamburg-Bergedorf, [p. 6]; Elsa Hirsch, Berlin, [p. 3]; Kurt Hirsch, Berlin, [p. 3]; Herwart von Hoff, Bremen, [p. 2]; Cuno Hoffmeister, Sonneberg, b. 1892, (28); Josef Hopmann, Leipzig, b. 1890, [p. 1], (9); Maria Hopmann, Sessig?, [p. 4]; Paul Hügeler, Berlin-Mariendorf,

b. 1892, [p. 5], (89); Hans Jensen, Kiel, [p. 5]; Jung (Oberbürgermeister v. Göttingen [mayor of G.]), [p. 1]; Albrecht Kahrstedt, Berlin-Dahlem, b. 1897, [p. 4], (12); Sigismund Kienast (?), Göttingen, [p. 7]; Hans Kienle, Göttingen, b. 1895, [USG director], (20); Harald von Klüber, Potsdam, b. 1901, [p. 6], (26); Hilde Knopf, Jena, [p. 4]; Otto Heinrich Julius Knopf, Jena, b. 1856, [p. 1], (6); Arnold Kohlschütter, Bonn, b. 1883, [p. 6], (24); Lore Kohlschütter, Bonn, [p. 2]; Arthur König, Jena, b. 1896, [p. 7], (94 ?); L. König, Jena [p. 7]; August Kopff, Berlin-Dahlem, b. 1882, [p. 5], (4); Margarethe Kopff, [p. 5]; E. Labitzke, Königsberg, [p. 2]; Paul Labitzke, Königsberg, b. 05.01.1890, [p. 1], (108); Johannes Larink, Hamburg-Bergedorf, b. 1893, [p. 4], (14); Leon Lichtenstein, Leipzig, b. 1878, (ident. not found); Inge-Lise Lillequist, Helsingfors, [p. 2]; Bertil Lindblad, Stockholm, SE, b. 1895, [p. 4], (81 ?); J. Lindig, Steinach/Thuer.; M. Lindow, Münster, b. 1880, [p. 3], (48); M. Linnemann, Göttingen, [p. 2]; Friedrich Wilhelm Hans Ludendorff, Potsdam, b. 1873, [p. 1], (3); Käthe Ludendorff, [p. 1]; Knut Lundmark, Lund, SE, b. 1889, (8); Sigmund Mauderli, Bern, CH, b. 1876, (59); Bruno Meyermann, Göttingen, b. 1876, [p. 6], (111); M. Meyermann, [p. 1]; H. Mollet, Olten, CH; Karoly Morawetz, Budapest, HU, b. 1899, [p. 2]; Josle (?) Müller, Potsdam, [p. 2]; Karl Müller, Wien, AU, b. 1866, [p. 4], (ident. not found); Rolf Müller, Potsdam, b. 1898, [p. 1], (100); Max Mündler, Heidelberg, b. 1876, [p. 5], (44); Axel Vilfred Nielsen, Aarhus, DK, b. 1902, [p. 6], (13); F. Pavel, Potsdam, [p. 5]; Cecilia Helena Payne, Cambridge, USA, [p. 4], (34); Johannes Peters, Berlin-Lichterfelde, b. 1869, (68); Karl Pilowski, Berlin-Dahlem, b. 24.08.1905, [p. 2], (40); Käthe Prager, Neubabelsberg, [p. 7]; Richard Prager, Berlin-Neubabelsberg, b. 1883, [p. 5], (1); Werner Rambauske, Stettin, [p. 2]; Göran M. Ramberg, Stockholm, SE, [p. 5]; Karl Reinmuth, Heidelberg, b. 1892, [p. 4], (81 (or 47?); H. Ritter, Berlin-Wilmersdorf, [p. 5]; [Mrs. Isaac-Roberts] Dorothea Roberts-Klumpke, Maison de la Rosa, Bonheur, FR, b. 1861, (51); Erwin Roux, Berlin-Friedenau, b. 1891, (62); Ilse Ruthe, Holzminden, [p. 2]; Kurt Ruthe, Holzminden, b. 1894, [p. 2], (43); Herzog Ernst von Sachsen-Altenburg, Wolfersdorf, Thuer., b. 1871, (5); H.-U. Sandig, Leipzig, [p. 6]; Carl Adam Wilhelm Schalén, Upsala, SE, b. 1902, [p. 5], (16); Werner Schaub, Leipzig, b. 1901, (79 ?); Otto Schlier, Heidelberg, b. 1902, [p. 4], (17); Josefa Schlüter, Osnabrück, [p. 3]; W. Schmitt, Bonn, [p. 6]; Heribert Schneller, Berlin-Babelsberg, [p. 6]; Erich Schönberg, Breslau, b. 1882, [p. 2], (29); Tyra Schönberg, Breslau, [p. 2]; Richard Reinhard Emil Schorr, Hamburg, b. 1867, (35); Else Schuler, Göttingen, [p. 6]; Max Schuler, Göttingen, [p. 6]; Walther Schulz, (Regierungsbaumeister, Stuttgart), [p. 3]; Walter Schulz (son of. Walther Schulz), Stuttgart, [p.3]; Agatha Schwarzschild, [p. 5]; Martin Schwarzschild, Göttingen, [p. 5], (32); Hildegard Seidel, [p. 1]; Heinrich Siedentopf, Jena, b. 1906, (15); Johann Wilhelm Jakob Anton Stein S.J., Roma, IT, b. 1871, [p.1], (75); Otmar Stenger, Göttingen, [p. 2]; Dagrun (?) Stenquist, Upsala, SE, [p. 5]; Erik Stenquist, Upsala, SE, [p. 5]; Bernhard Rudolf Sticker, Bonn, b. 1906, [p. 6]; 47; Joachim Stobbe, Kiel, b. 1900, [p. 5], (42); H. Strassl, Göttingen, [p. 3], (50); Hermann Strebel, München-Herrsching, b. 1868, [p. 2], (63); Bengt Georg Daniel Strömgren, Copenhagen, DK, b. 1908, [p. 4], (36); Svante Elis Strömgren, Copenhagen, DK), b. 1870, [p. 1], (7);

Elisabeth Stumpff, Breslau, [p. 5]; Karl Stumpff, Breslau, b. 1895, [p. 5], (30); Laura Stumpff, Hamburg, [p. 5]; Ludwig Terkßn, Budapest-Savabhegy, HU, b. 1877, (58); Walter Theile, Marburg, [p. 2]; Bruno Thüring, Muenchen, b. 1905, [p. 2], (27); Valentiner (?), Kurator der Georgia Augusta, Göttingen, [p. 4]; Carl Vick, Hamburg, b. 1875, [p. 6], (77); Walter Augustin Villiger, Jena, b. 1872, (64); Julie Marie Vinter Hansen, Copenhagen, DK, b. 1890, [p. 4], (49); Heinrich Vogt, Heidelberg, b. 1890, [p. 3], (101); A.A. Wachmann, Hamburg, [p. 2]; Lotte Waschmann (?), Göttingen, [p. 2]; O. Wachtl, Bonn, [p. 2] ("W/8"), (79 ?); Kurt Walter, Königsberg, b. 22.09.1905, [p. 7], (81 ?); Josef Weber, Leipzig, b. 1888, (78); Johannes Wempe, Göttingen, [p. 7] ("25 h WTZ"), (10); Dorothea Werner-Starke, Jena), [p. 6]; Helmut Werner, Jena, [p. 6]; Rupert Wildt, Göttingen, b. 1905, [p. 7], (33); Alexander Wilkens, Muenchen, b. 1881, [p. 2], (11); F. Wunschmann, Potsdam, [p. 5]; Ilse Wunschmann, Potsdam, [p. 5]; Walter Zimmermann, Görlitz, b. 1876, (ident. not found).

Cecilia's signature is still to be seen in the guestbook she signed at the meeting of the AG [Fig. 35].

Arthur Eddington [Fig. 36] was an important mentor of Cecilia's, and a man whom she admired greatly. She first encountered him at Cambridge, and they met at a number of astronomical meetings during the 1920s and 1930s.

The meeting of the Astronomisches Gesellschaft was filled out with excursions and parties. One was an outing on the River Weser at which Cecilia and her mentor found time to be photographed together [Fig. 37].

4. MARRIAGE, FAMILY AND ASTROPHYSICS

In 1934, Cecilia and Sergei were married in New York City, with the full support of others at the Observatory. A myth has arisen that they eloped, but their decision to marry was known and approved. They embarked on an ambitious honeymoon to the west, returning via the Panama Canal on the President Pierce. On their auto trip driving west, they stopped at the Flagstaff Observatory and met there with the brothers E. C. and Vesto M. Slipher [Fig. 38].

Once in California, they spent time in Pasadena, home of the California Institute of Technology and Mount Wilson Observatory offices. Here Cecilia [Fig. 39] is captured on one of the palm-lined streets of what was at that time a quiet, sunny town.

One evening the newly-weds dined in a Pasadena restaurant with other astronomers [Fig. 40]. From the left, Walter Baade, CPG, -, Carol Angers Reike (?), -.

Back in Massachusetts, Cecilia and Sergei renewed their friendship with Frances Wright and her frequent escort Otis Scammon and participated in outings to Oak Ridge in Harvard, Mass. where telescopes were located away from the city lights of Cambridge. On this occasion at Oak Ridge [Fig. 41] they are joined by Professor and Mrs. Pannekoek. Professor Pannekoek was an astronomer and historian of astronomy.

Cecilia continued her association with Professor Russell from Princeton, and in 1935 the Russells visited the Gaposchkins at their home on Lexington Ave. in Cambridge [Fig. 42]. From the left: Mrs. Russell, Cecilia, Prof. Russell and Meghnad N. Saha.

In 1936, the couple were captured in this photo in Sergei's office, which was on the ground floor of Building C. [Fig. 43].

The Gaposchkins had moved to Lexington, Mass. and in 1938, Cecilia and her children received a visit from Frances Wright [Fig. 44]. From the left; Frances Wright, Katherine, Edward and Cecilia.

The Gaposchkins attended the meeting of the American Astronomical Society in Philadelphia, Pennsylvania in 1937, where Cecilia and Otto Struve were photographed walking between sessions [Fig. 45].

As World War II began to erupt, many talented astronomers were uprooted from their homes. Some found temporary refuge and the opportunity to continue their work in astronomy at the Harvard College Observatory. They were caught in this group photo on the steps of Building D [Fig. 46]. Back row, from left: George Z. Dimitroff (Bulgarian), CPG (English), Luigi Jacchia (Italian); Middle row: Donald MacRae (Canadian), Zdenek Kopal (Czech), Richard Prager (German), Sergei Gaposchkin (Russian); Front row standing: Shirley Patterson (Canadian), Maria Paris Pishmish (Armenian/Turk), Odon Godard (Belgian); Front row seated: Bart J. Bok (Dutch), Jaakko Tuominen (Finn), Masanki (?) Huruhata (Japanese), Luis Erro (Mexican)

Another long time friend and an astronomer much admired by Cecilia was Ejnar Hertzsprung, here [Fig. 47] seen at his desk holding an eyepiece and glass photographic plate.

Though war was looming, an exciting conference on novae and white dwarfs was arranged in Paris, in July1939 and the Gaposchkins attended [Fig. 48]. Back row from left: C. S. Beals, B. Edlén, P. Swings, B. Strömgren, S. Chandrasekhar, W. Baade. Middle row: G. P. Kuiper, K. Lundmark. Front row: F. J. M. Stratton, C. Payne-Gaposchkin, H. N. Russell, A. Schaler (conference organizer), A. S. Eddington, Sergei Gaposchkin (photo courtesy of S. Chandrasekhar).

A couple of years later, Cecilia and her family were photographed in the garden of their home in Lexington [Fig. 49] with visitors Martin Schwarzschild and Otis Scammon. The photo was probably taken by Frances Wright. From the left, are Katherine, Martin Schwarzschild, Cecilia, Otis Scammon, Peter, Sergei and Edward.

On another occasion, Cecilia is seen in a flowered dress in her garden [Figure 50].

Early in 1941, Cecilia was featured in an article about her unusual work as an astronomer in the Boston Sunday Post, published January 4, 1942. The article was illustrated with a montage of several photos that captured what she did. [Fig. 51] shows her at the telescope and [Fig. 52] examining a plate with an eyepiece.

In February that same year, the Gaposchkins traveled to Mexico with many other astronomers for the opening of the Tonantzintla Observatory in Mexico. Cecilia is visible in a hat at the left edge of the photo [Fig. 53], with just the cheek and chin of Sergei visible beside her. The President of Mexico, Avila Camacho, is seated at the table, with Harlow Shapley two people to his right.

In 1946, Cecilia is seen with her family in her office [Fig. 54] (from left, Katherine, Edward, Sergei and Cecilia) and operating the blink microscope for the detection of variable stars [Fig. 55].

In 1948, the Gaposchkins took a trip across the US to visit the western observatories, and in 1949, Cecilia and Katherine traveled to Europe, where Fig. 56, with the look of "the vision splendid," was taken. Donald Menzel is shown at Sacramento Peak in 1953 in Fig. 57.

In 1956 Sergei went with Peter to observe at Mt. Stromlo, Canberra, in Australia. That year the Gaposchkins sent out this stellar Christmas card with the motto, "The stars never set on the Gaposchkins" [Fig. 58].

Times changed at the Observatory, and in 1957 Cecilia was appointed Chair of the Department of Astronomy. The Harvard Crimson wrote a story about her, published Feb. 23, 1957. In 1958, on the occasion of the meeting of the International Astronomical Union that was held in Moscow, USSR, the Gaposchkins visited the Soviet Union. This was the first time Sergei had returned to his homeland since he left in 1922. In Fig. 59 Cecilia and Sergei are looking at an instrument and Sergei holds a copy of the General Catalogue of Variable Stars. (Courtesy of B. Vorontsov-Velyaminov.)

During the last decade of her career, Cecilia received numerous awards and honorary degrees. She particularly valued the Rittenhouse Medal of the Franklin Institute located in Philadelphia, Pennsylvania, which honor she received in 1961 [Fig. 60].

At about this same time she was invited to be on a panel [Fig. 61] with the grandson of Charles Darwin, a noted physicist. In a somehow fitting "full circle", she connected again with a member of the Darwin family, after many decades of contributing to her own world of science.

In 1973, the Gaposchkins traveled to Australia for the 15th General Assembly of the International Astronomical Union at Sydney. Sergei sketched profiles of a few participants, including two of Cecilia [Fig. 62].

Finally, her characteristic signature is shown in Fig. 63. She was a person of strong will, stormy temperament, love for others and tremendous love for science. She never lost her belief that "Nature never did betray the heart that loved her".

5. A WORD ON PRONUNCIATIONS

As it is difficult to convey correct pronunciations in print, I have provided some little verses that may help clarify how to say Cecilia (rhymes with 'lobelia' - 'Sesseelya) and Helena (rhymes with 'bell-in-a), Sergei (rhymes with 'fair play') and Gaposchkin (rhymes with 'oh gosh!' kin). Sergei used the German spelling of his last name (with 'sch) until his 70s when he changed it to 'Gaposhkin'.

Cecilia

A starry-eyed lady Cecilia
Fell in love with a clump of lobelia
She found her Sergei
Who wanted fair play
And they romped through the clump of lobelia!

Helena

Our heroine also called Helena
Was stargazing when her glance fell in a
Cluster of stars
Full of multiplet bars
That sagacious young lady called Helena.

Gaposchkin

A traveling pair, the Gaposchkins
Were rarely alone so odds-bodkins,
They vanished one day
In a fabulous way
Leaving family to wonder, Oh gosh,kins!

ACKNOWLEDGMENTS

I am indebted to the efforts of many people who assisted me in identifying people and locations in several of the photographs in this article. Most especial thanks to Axel Wittmann and Willi Deinzer who made a heroic effort to identify the individuals at the 1933 AG meeting in Göttingen, Germany and to Michael Saladyga of the AAVSO who assisted with other photos. Sergei Haramundanis, Nancy Evans and Barbara Welther were also most helpful.

REFERENCES

Catch, J. R. 1990 A Buckinghamshire Polymath: Edward John Payne, Records of Buckinghamshire, Vol 32, 120

Hill, W. H, Hill, A. F. & Hill, A. E. 1902 Antonio Stradivari, his Life and Work

Jeffreys, Lady B. 1985, letter to KH, 11 August

Payne, E. J. 1892, 1899 History of the New World Called America

Vierteljahresschrift der AG, 1935 Vol. 68, Heft 4, 295

Fig. 1. Cecilia at about 1, Wendover

Fig. 2. From left to right, Cecilia, Mrs. Hill, Emma Payne, Humfry, Mr. Hill and Edward Payne

Fig. 3. Cecilia and Humfry Payne, 1904

Fig. 4. Cecilia and Humfry Paync, c. 1908, in theatrical costume

Fig. 5. Emma (mother) and Cecilia - 1910

Fig. 6. Holywell Lodge, c. 1910, Wendover

Fig. 7. Wendover School, photo 1994

Fig. 8. Dora Pertz, photo c. 1890

Fig. 9. Lenora, Humfry and Cecilia - 1912

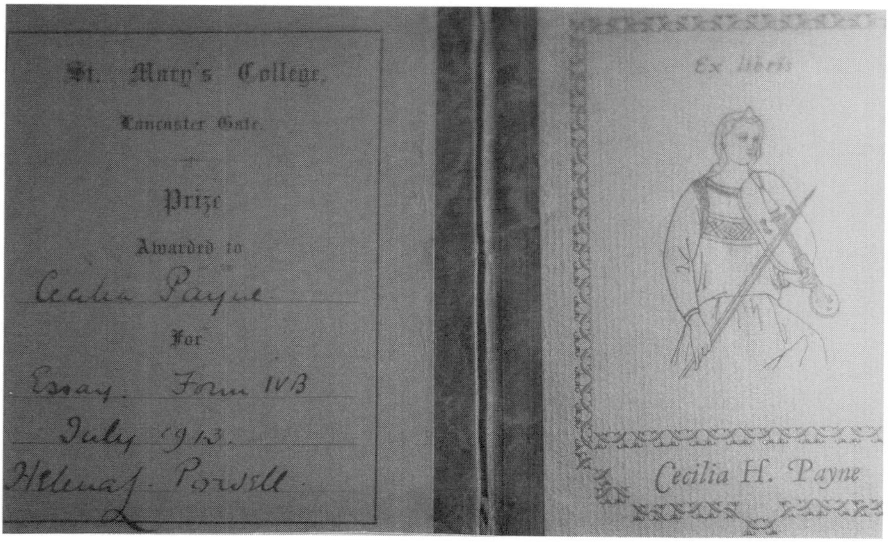

Fig. 10. School prize and Cecilia's bookplate - 1913

Fig. 11. Studio portrait - labeled Swain's, c. 1920

Fig. 12. Newnham College Gardens - 1994

Fig. 13. Newnham Observatory

Fig. 14. Rutherford and Bohr

Fig. 15. E. A. Milne, courtesy of Astron. Society of the Pacific

Fig. 16. C. Payne, in local costume, 1923

Fig. 17. Meeting of the AAVSO at HCO, Oct. 13, 1923

Fig. 18. Meeting of the AAVSO at HCO, Oct. 11, 1924

Fig. 19. Meeting of British Association for the Advancement of Science, Toronto, 1924

Fig. 20. CP on grounds of University of Toronto

Fig. 21. HCO 'computers' - 1925

Fig. 22. CP in hat with green ribbon - 1925

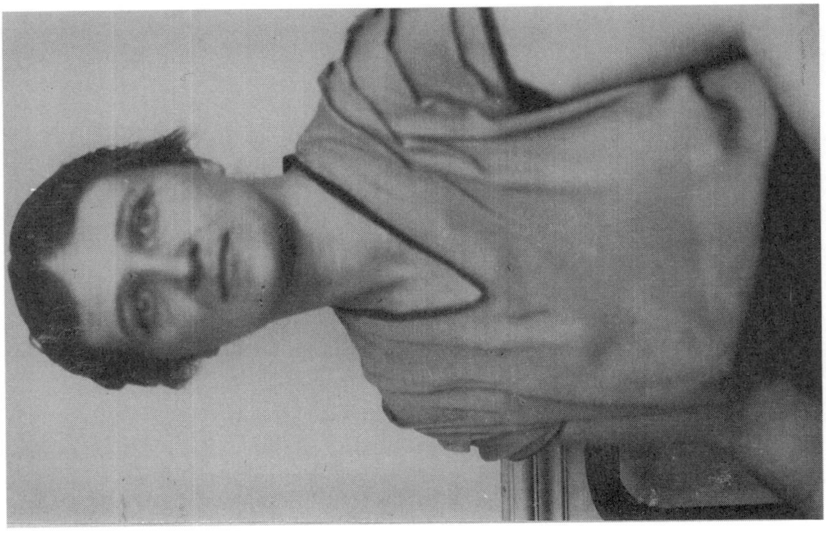

Fig. 23. CP - slightly formal picture - 1925

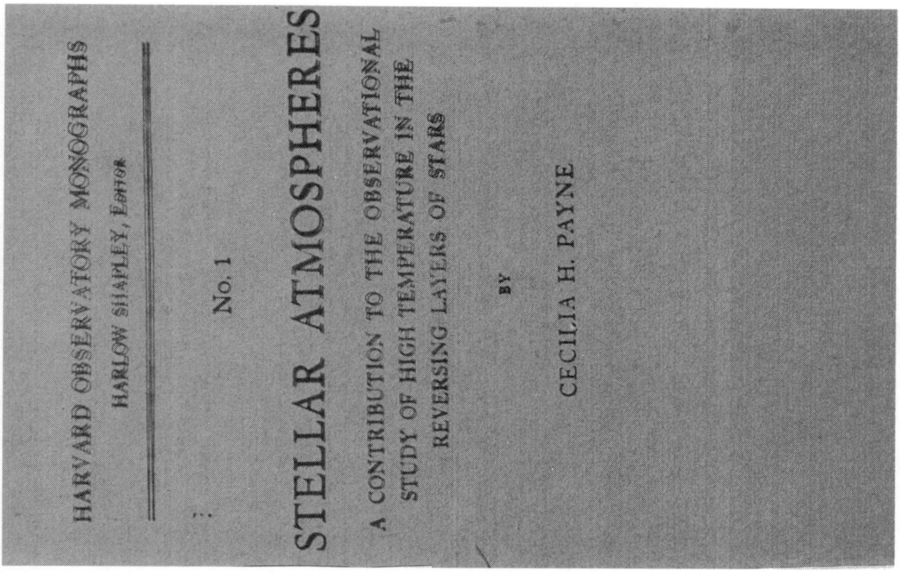

HARVARD OBSERVATORY MONOGRAPHS

HARLOW SHAPLEY, EDITOR

No. 1

STELLAR ATMOSPHERES

A CONTRIBUTION TO THE OBSERVATIONAL STUDY OF HIGH TEMPERATURE IN THE REVERSING LAYERS OF STARS

BY

CECILIA H. PAYNE

Fig. 24. Title page of Thesis, 1926

Fig. 25. Observatory Pinafore, December, 1929

Fig. 26. CP and Frances Wright, c. 1929/30

Fig. 27. CP, FWW and their auto - 1930

Fig. 28. CP on mule at Grand Canyon - 1930

Fig. 29. Frances W. Wright on mule, 1930

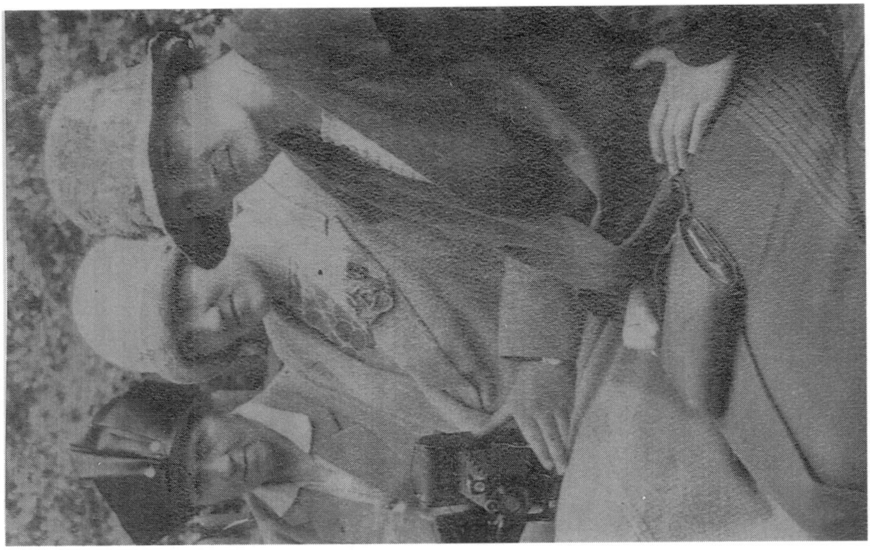

Fig. 30. UK - 1933 (?) or earlier. Emma, Leonora and Cecilia

Fig. 31. Boris Petrovich Gerasimovich, Pulkova, 1933

Fig. 32. Pulkova, 1933

Fig. 33. Meeting of the Astronomisches Gesellschaft, 1933

Fig. 34. Labeled AG group photograph

Fig. 35. Signature of Cecilia Payne, 1933

Fig. 36. Arthur Eddington, 1933

Fig. 37. Cecilia and Arthur Eddington

Fig. 38. CPG with brothers, Slipher, Flagstaff, 1934

Fig. 39. CPG, Pasadena, 1934

Fig. 40. Dinner party, Pasadena, 1934. Walter Baade, Cecilia, ?, Carol Angers Reike, ?, ?

Fig. 41. With Prof. and Mrs. Pannekoek, Oak Ridge, 1935

Fig. 42. Mrs. Russell, Cecilia, H. N. Russell, M. N. Saha

Fig. 43. Office of Sergei, with SG, Nov. 1936

Fig. 44. Summer 1938, Lexington. Francis Wright, Katherine, Humfry and Cecilia

Fig. 45. Cecilia and Otto Struve, AAS meeting, Ann Arbor, 1938

Fig. 46. "Foreign" Research Group at HCO, July, 1939

Fig. 47. Ejnar Herztsprung (Courtesy Astronomical Society of the Pacific)

Fig. 48. Conference on Novae and White Dwarfs, Paris, 1939

Fig. 49. 1941, Lexington, Massachusetts. Katherine, Martin Schwarzschild, Cecilia, Otis Scammon, Peter, Sergei and Edward

Fig. 50. CPG in flowered dress, 1941

Fig. 51. Boston Sunday Post, CPG at telescope, 1942

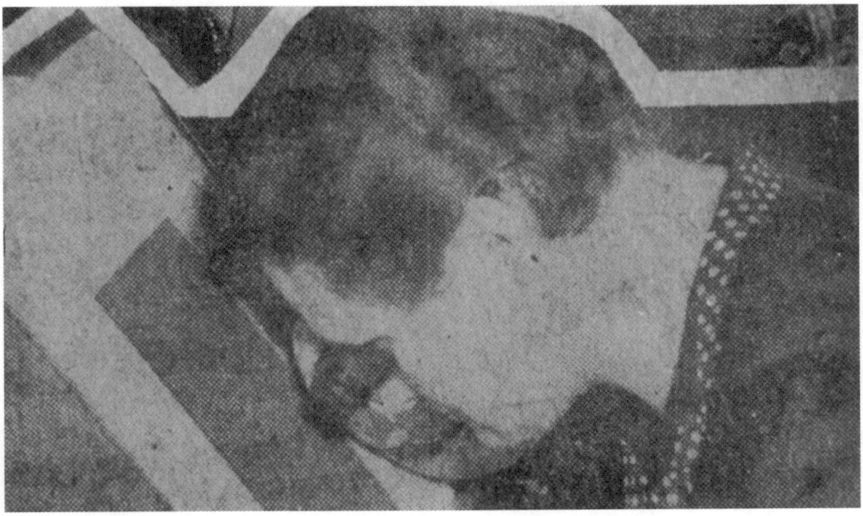

Fig. 52. Boston Sunday Post, CPG with eyepiece

Fig. 53. Opening of Tonantzintla Observatory, 17 Feb. 1942

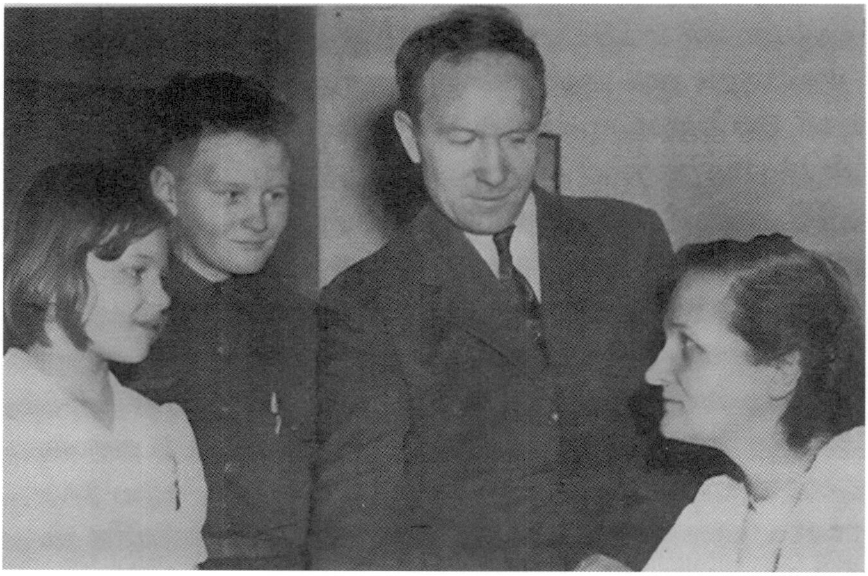

Fig. 54. In her office on the top floor of Building D, 1946. Katherine, Edward, Sergei and Cecilia

Fig. 55. Operating the Blink Microscope, c. 1946

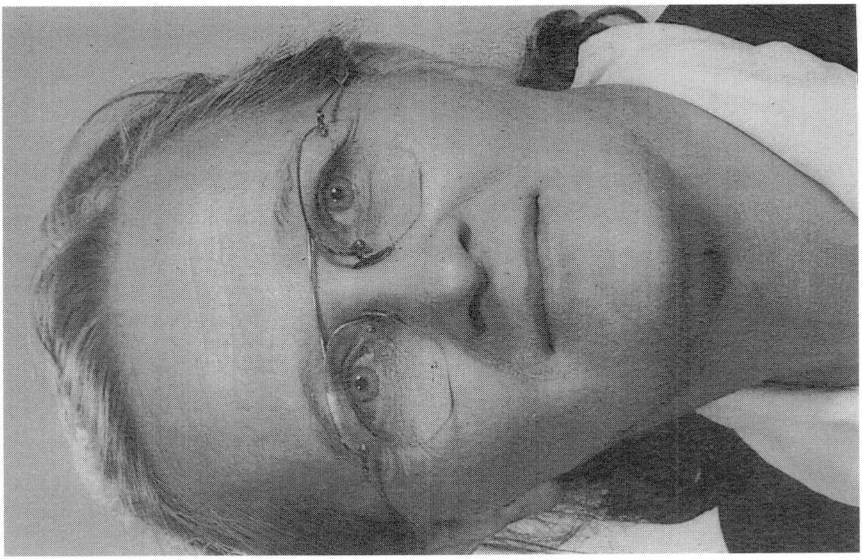

Fig. 56. Cecilia "the vision splendid", 1949

Fig. 57. Donald H. Menzel on Sacramento Peak, NM, 1953

Fig. 58. Christmas Card, 1956 - 57

Fig. 59. At the IAU meeting in Moscow, 1959. Cecilia and Sergei

Fig. 60. Receiving Rittenhouse Medal, Franklin Institute, 1961

Fig. 61. With Sir Charles Galton Darwin - 1961 (?)

Fig. 62. Participants of IAU General Assembly, 1973

Fig. 63. CPG signature

NEW DIRECTIONS IN VARIABLE STAR RESEARCH

Janet Akyüz Mattei & Kerriann H. Malatesta

American Association of Variable Star Observers (AAVSO), Cambridge, Massachusetts

Abstract: As we joyously celebrate the centenary of Cecilia Payne-Gasposchkin's birth, we look at the contribution this remarkable woman has made to astronomy, particularly her contribution to variable star astronomy. Also discussed are new directions in variable star astronomy since Cecilia's reign, including issues that have risen and recommendations for the future of variable star surveys and databases worldwide.

1. PERSONAL RECOLLECTIONS BY ONE OF US (J. A. Mattei)

My first encounter with Cecilia Payne-Gaposchkin occurred when I was a senior in high school in Izmir, Turkey. During my studies that year, I had the good fortune to get acquainted with Cecilia through her wonderful book titled "Introduction to Astronomy". This informative book was the required text for the astronomy course that I was enrolled in at the American Collegiate Institute - one of the five American high schools for Turkish students. The eloquently written text had a profound effect on me. Not only did I find the scientific content to be fascinating, but I was also impressed by beautiful poetic quotations at the beginning of each chapter. The information contained in the book built the foundation of my knowledge and interest in astronomy. At the time, I never thought that I would ever have the opportunity to meet Cecilia.

However, in 1972, ten years after my initial introduction to this inspiring woman, I had the privilege of meeting Cecilia when I started working at the American Association of Variable Star Observers (AAVSO). On several occasions I took the time to attend her presentations at Harvard College Observatory. Her talks were superb. I was extremely impressed with the breath of knowledge she had about variable stars, how she could list one variable star after another by name and immediately cite their unique characteristics. It was as if each variable star was her friend and she knew each of their personalities by heart. Another unforgettable trait of her talks was that she rarely looked at the audience. Instead her sight was turned skyward in the direction of the stars. It was as if she were conversing with them while the audience eavesdropped on the one-way conversation. In a unique way, she conveyed the sense of wonder and enthusiasm that she felt to her audience.

Later, when Cecilia was writing "Stars and Clusters", I had the opportunity to get to know her better. In her book she used a good number of AAVSO light curves, and also used "The AAVSO Variable Star

The Starry Universe: The Cecilia Payne-Gaposchkin Centenary
A. G. Davis Philip & Rebecca A. Koopmann, eds.
p. 67-76

Janet Akyüz Mattei
AAVSO
25 Birch St., Cambridge
Massachusetts 02138-1205
JMATTEI@AAVSO.ORG

Atlas" for finder charts. During this time we had a chance to get together and discuss variable stars on several occasions. With each meeting I was further filled with admiration and awe at the depth and breadth of her knowledge of astronomy. I was also very appreciative of the fact that although I was just a novice in the field of variable stars she treated me as a friend and a colleague. I will always treasure my memories of Cecilia

2. CECILIA AND VARIABLE STARS

2.1 The Early Years

Cecilia's inquisitive nature and pursuing mind blossomed as a child. She was interested in many disciplines, but favored the natural sciences. At the age of five Cecilia reportedly saw a meteor whereupon she expressed her desire to become an astronomer. In childhood she expressed the concern that if she did not grow up fast enough, then there would be no research for her to do and thus, she must get started immediately (Payne-Gaposchkin 1996).

One of Cecilia's first encounters with variable stars came when a good friend, L. H. Comrie of New Zealand, gave her Caroline Furness' book on the subject while in Newnham College at Cambridge University "to guide her first observations" (Kidwell 1996). During her university studies at Cambridge, Cecilia first explored the fields of botany, chemistry and physics. However, after hearing A. S. Eddington speak of his eclipse expedition to the Island Principe and implication of the results, particularly to general relativity, "she resolved to become an Astronomer". In 1923, after graduating from Newnham College, she came to Harvard College Observatory (HCO) on a Fellowship to do research under Harlow Shapley, the new director of HCO. While working with projects involving stellar magnitude sequences, and later with variable stars, she started her graduate work in stellar atmospheres - a field that intrigued her. Withing two years, Radcliff College/Harvard awarded her the first Ph.D. in Astronomy ever granted to a student at HCO.

Fig. 1. Cecilia Payne-Gaposchkin, 1938. Courtesy of AAVSO

Although Cecilia was well on her way to making major advances in astronomy, her love of all of the natural sciences never faded. In a remark about her early years spent at HCO, this sentiment is reflected, "I had spent my first few years there [Harvard] in ranging over the astronomical photographs, collecting and classifying celestial flora. I choose the word deliberately, for my approach was that of the field naturalist. Just as I had learned the intricacies and pitfalls of systematic botany 15 years earlier, I had spent almost a decade in identifying and pigeon-holing the astronomical species" (Payne-Gaposchkin 1996, p. 182).

2.2 The Milton Variable Star Bureau

Harvard College Observatory, with its world-renowned collection of photographic plates, was the center of photometric and spectroscopic studies and classification during the directorship of Edward C. Pickering at the close of the 19th century and the beginning of 20th century. Pickering had collected a broad base of visual and photographic photometric data and had begun a systematic study of variable stars. However, due to lack of funds, the variable star project was discontinued. The next director of HCO, Harlow Shapley, continued the collection of the HCO photographic plates with a special interest in pursuing a "survey of variable stars as an approach to the structure of the Milky Way" (Payne-Gaposchkin 1996). The photographic plates, as it turns out, served both in variable star discovery and in the study of these stars on a large scale.

In the late 1930s, Cecilia revitalized the variable star project that Pickering had initiated. The time was right for a photographic survey - for an arrangement of variable stars in a coherent scheme. Variable stars were seen to have a great variety of properties. No longer were they considered "pathological stars". Cecilia realized that there was a clear and definite domain which could be seen by studying a variety of stellar properties. As such, she proposed a project to study all known variable stars brighter than a photographic magnitude of 10 on Harvard patrol plates. As a result, the Milton Variable Star Bureau was established in 1937, with Cecilia and her husband Sergei Gaposchkin in charge of the program. The survey, which was delayed due to World War II, was completed in 1953. It involved photographic plates ranging in date from 1898 to 1947 and 1,263,562 magnitude estimates for 1512 variable stars. For this project, the sky was divided into 54 fields, with seven zones each, in increments of 30 degrees in declination.

The Gaposchkin's classified the variables into the following 11 major variable star categories: Long Period (Mira), Semiregular, Irregular, RV Tauri, Cepheid , RR Lyrae, Eclipsing binaries, R Coronae Borealis, U Geminorum and Z Camelopardalis and Novae. The Milton Variable Star Bureau project was the most ambitious, complete, and thorough survey to have taken place in the first half of the 20th century. The results from their studies were published in the "Annals of the Harvard College Observatory": fields 28 to 54 in "Harvard Annals" 115 (Payne-Gaposchkin et al. 1952a), and fields 1 to 27 in "Harvard Annals" 118 (Payne-Gaposchkin et al. 1952b). In Dorrit Hoffleit's recent and extensive review (2000) of this impressive project, she notes that the Gaposchkin's questioned the variability of only 103 stars in their survey. Hoffleit compared these stars with the entries in the "General Catalogue of Variables Stars" (GCVS) (Kholopov 1985 - 1987, Samus 1990) and reported that 36 have now been confirmed as variables classified by type), 21 have "modern classification", 16 are discordant with GCVS, and the rest are still "dubious". Thus, these stars are good candidates for further study.

2.3 Further Contributions

Cecilia and Sergei presented their preliminary Milton Bureau-based findings in a book they published jointly in 1938 titled "Variable Stars", which was one of the first books to contain global observational knowledge of variable stars. Additionally, although Cecilia was considered to be an expert on different types of variable stars, she was particularly interested in novae, or exploding stars. Through her work with variable stars, she contributed a great deal to the model of the nova as an exploding star. Her book, "Galactic Novae", published in 1957 continues to be an icon in nova studies nearly 50 years later.

In the 1950s and early 1960s, another major contribution of Cecilia and her husband Sergei had been the undertaking of the study of the Large and Small Magellanic Clouds, again using the Harvard photographic plates, which involved nearly 2,000,000 observations. This survey resulted in the discovery of 3000 to 4000 variables in the two clouds. More extensive information on this survey is given in Kenneth Jane's paper in this publication.

In the 1970s, with the advent of satellite technology, Cecilia predicted that far more information could be learned by studying variable stars from gamma ray all the way to radio wavelengths. Variable star research has indeed taken an impressive leap through multiwavelength research: we can now correlate ground- and satellite-based observations, and as a result, we are learning more and more about the nature of and the physics behind many of the sky's enigmatic objects, particularly exploding stars and transient events.

One of Cecilia's final contributions was presented in the "Annual Review of Astronomy and Astrophysics" paper titled "The Development of our Knowledge of Variable Stars" (1978). In it, Cecilia summarized how our knowledge of variable stars accumulated in the twentieth century and how what was once held sacred about variable stars in 1918 by the world-renowned astronomer, Henry Norris Russell, has proven to be otherwise. In this paper, Cecilia debated five points:

A) That variable stars could be classified into natural groups. While this is indeed true, Cecilia showed that the variable star groups far exceeded the beliefs of the early 20th century, with Cecilia having played a key component in the extension of the classes of variable stars. Again referring to her background in botany, she writes "...the classification of variable stars has undergone a change that recalls the supersession of the Linnaean system by the modern system of botanical classification" (Payne-Gaposchkin 1996).

B) The belief that almost all variable stars are giants. As Cecila points out, this conclusion is no longer applicable.

C) That Cepheid variables alone provide evidence that stellar evolution is extremely slow, and hence cosmological processes occur over a long time scale. Cecilia notes that we know this to be true, however this conclusion is not drawn from the class of Cepheid stars alone

D) That Russell's mechanism for novae outbursts was on the right path, but the specific details were not as he expected.

E) That we know very little about stellar variation. While there are still many unknowns, Cecilia comments with spirit that "happily this is no longer true..".

It is undeniable that Cecilia had a dramatic impact on the status of variable stars. She is quoted as saying: "When I first became an astronomer, variable stars were regarded as a sort of second-class problem. Perhaps I have had a small hand in making them respectable" (Payne-Gaposchkin 1996). Through her unique ability in correlating and systemizing large quantities of information, and bringing together previously unrelated facts and seeing a pattern in them, her findings not only led to a much better understanding of how and why stars vary in their brightness, but she also provided important clues about the structure of the Milky Way and other galaxies.

3. NEW DIRECTIONS IN VARIABLE STAR ASTRONOMY

By the end of the 19th century, about 1000 variable stars were known to exist. By stark contrast, the close of the 20th century shows that the previous sum is far eclipsed by the present count with 35,985 known (Kazarovets et al. 2000) and 14,810 suspected variable stars (Kholopov 1982). In addition, there are tens of thousands of yet uncataloged stars that are found to be varying as a result of several surveys, such as searches for gravitational lensing, supernovae and gamma ray burst afterglows. What has had such a dramatic impact on counts? Technology. The technology to scan the entire sky or even a portion of the heavens with automated telescopes or satellites has advanced the numbers so greatly that it is now becoming a challenge to keep up with the demands of naming and "pigeon-holing" these new found variable stars. What follows are a few examples of surveys with satellites and and ground-based telescopes.

4. SURVEYS WITH SATELLITES

In the last three decades, starting first with the X-ray satellite UHURU in the early 1970s, there have been many satellite surveys from X-ray to infrared wavelengths that have scannned the sky. These searches have included the X-ray satellite High Energy Astronomical Observatory-2 (HEAO-2) in late 1970s, the European X-ray Observatory Satellite (EXOSAT) and the infrared satellite (IRAS) in early 1980s, followed by a more sensitive X-ray satellite, the Rontgen Satellite (ROSAT), as well as the Extreme Ultraviolet Explorer (EUVE), and Infrared Satellite Observatory (ISO). The general mission of these surveys was to expand our knowledge of the universe beyond the optical wavelengths, and to discover new objects or study those already known to exist in multiwavelength regimes. In the process, many new types of objects, such as X-ray binaries and other variables, have been and continue to be

(Hipparcos), which surveyed the whole sky many times during the period of 1989 - 1993, has not only provided the parallax and proper motion for 118,218 stars with unprecedented accuracy, but it has also measured the brightness of these stars over the lifetime of the mission with milimagnitude accuracy. The acquired photometric measurements led to the discovery of 11,597 variable stars brighter than magnitude 12, with 71 % as new variables. These variables have been handled in an examplary fashion by the Hipparcos Photometry team. Mainly, the satellite data were analyzed sufficiently so that the results for most of the new variables satisfied the GCVS criteria; enough to render formal cataloging and naming of these stars in the 74th Special Name List of Variable Stars (Kazarovets et al 1999). In addition their light curves were published in Volume 12 of "The Hipparcos and Tycho Catalogues" (ESA 1997).

Tycho, an auxilary instrument of the Hipparcos satellite star mapping equipment, measured the brightness of more than 1 million stars with a lesser but still unprecedented accuracy. The Tycho observations have been published in "The Hipparcos and Tycho Catalogues" (ESA 1997) and have been recently revised and published in "The Tycho-2 Catalogue". Tycho photometry, however, has not yet been sufficiently analyzed for formal results on variable stars. It is expected that when the data receive a thorough analysis, thousands of new variables down to a limiting magnitude of 11 will be discovered.

As a follow up to Hipparcos, the Global Astrometric Interferometer Astrophysics (GAIA) satellite is scheduled to be launched by 2012. During its expected seven years of operation, it will undoubtedly make a dramatic impact in all areas of astronomy, and will surely revolutionize the field of variable star astronomy. The all-sky survey will include 1 billion stars, covering a sky density of about 25,000 stars per square degree, with a limiting magnitude of 20, viewed in 4 broad and 11 medium bands. Never before has there been such a homogenous, all-sky survey to a limiting magnitude of 20.

In the coming years, several more satellites will be launched. Some have a mission to search for planets around stars, while some will study microvariability and oscillations of stars. Thus, the coming decades will provide fertile ground for variable star discovery. Below we give the names of a sample of these satellites. Although from their abreviated name it is hard to guess their mission, their full name is quite descriptive: Convection, ROtation, and planetary Transits (COROT), Microvariability, Oscillations of STars (MOST) and Measuring Oscillations in Nearby Stars (MONS).

5. GROUND BASED SURVEYS

Along with satellite surveys, there have been an impressive number of ground-based surveys in the last decade, mostly with a mission of finding gravitational lensing events. Some of these surveys are: Massive Compact Halo Objects (MACHO), Optical Gravitational Lensing Experiment (OGLE), and Experience pour la Recherche d'Objets Sombres (EROS). Although variable star studies have not been the primary goals of these missions, the accumulated photometry from these surveys has provide a gold mine of information in terms of the discovery and study of variables stars. Unfortunately, some of these projects, such as the MACHO survey, have been terminated due to the lack of funding.

Table 1. Sample Sky Surveys

Survey	Band	Limiting Mag.	Area Covered	Stars Surveyed	New Variable Found
ASAS	I	I = 13.0	300 Sq deg In 50 fields	> 140,000	3733
FASTT	Wide R	R = 17.8	389 sq. deg.	1,268,732	4300
KAIT	Unfiltered CCD	~19.0	~1000 gal. per night	~1000 gal. per night	>100
MACHO	Nonstd R,B	R ~ 16	Galactic bulge. & Mag. Clouds	30,000,000	Tens of Thousands
OGLE	I	I = 14-18	Galactic bulge & Mag. Clouds	Hundreds of Thousands	2861
PLANET	Primarily I	I = 19	5x5 arcmin fields at 20-5 fields/yr	2000-15000 stars/field	~0.5% per Field
ROTSE	Unfiltered broadband CCD	15.5	Entire sky	Millions	1600
STARE	B,V,R broad-band filters	R = 8 - 14	Constellations of Auriga, Perseus, Pegasus, Leo Minor and Bootes	8000-50000 stars/field	Several % per Field
TASS	Mainly V, I and some R	V = 13.0, R = 13.0, I = 12.5	2160 sq, deg im strip of dec. from +1 to - 5 deg.	367,241	60

Deciphering the alphabet soup of surveys: All Sky Automatied Survey (ASAS), Flagstaff Astrometric Scanning Transit Tesescope (FASTT), Katzman Automatic Imaging Telescope (KAIT), Massive Compact; Halo Object (MACHO), Optical Gravitational Lensing Experiment (OGLE), Probing Lensing Anomalies NETwork (PLANET), Robotoic Optical Transient Search Experiment (ROTSE), Stellar Astrophysics & Research on Exoplanets (STARE), and the Amateur Sky Survey (TASS).

The search for gravitational lensing, gamma ray burst afterglows and supernovae have been the main mission of most of these surveys. In each case, however, the discovery of other variables have been a natural by-product. In Table 1, we list a sample of automated telescope surveys mentioned above and give some characteristics of each, including the number of variables discovered as of October 2000.

6. THE AAVSO, VARIABLE STAR CHALLENGES AND RECOMMENDATIONS

The American Association of Variable Star Observers (AAVSO),which was part of the Harvard College Observatory and became independent in 1954, has a mission to coordinate, compile, process and disseminate observations made (mostly) by amateur astronomers to the astronomical community. With a digitized database of over 10 million observations for 3500 (mostly) large amplitude variables, it has the world's largest database of variable stars. Variable star observers, mainly amateur astronomers, have been the key in the monitoring of these stars for close to a century, following the intial dream of its founders, William Tyler Olcott and Edward C. Pickering in 1911. It is through careful monitoring over decades that it is possible to record and witness longterm and often, the unusual behavior of variable stars which, in turn, makes it possible to understand their true nature. Fig. 2, on the next page, shows longterm AAVSO lightcurves of some Mira variables (W Aql, R Lep, Y Per and RV Cen). These examples show that without the decades of observations contained within AASVO International Database, it is not possible to study the longterm behavior and the unique features that arise in these stars. It is interesting to note that with the technological advances in the space research of variable stars the role that variable star observers and the AAVSO representing them plays has become vital in scheduling satellite and ground-based observations. This amateur-professional collaboration allows for simultaneous observering during multiwavelength observing runs, and is particularly important in correlating multiwavelength data with the optical data. In the last three decades, the observations of AAVSO observers have helped with observing runs and research of variable stars with 24 satellites. Just in the last year, AAVSO observations have assisted observing runs with ground-based telescopes or instruments such as Palomar Testbed Interferometer, Very Large Base radio telescope, William Herschel telescope, telescopes at the European Southern Observatory, and space satellites such as Rossi X-ray Timing Explorer (RXTE), Hubble Space Telescope (HST), Far Ultraviolet Satellite Explorer (FUSE), and Chandra X-ray satellite.

Variable star observers today have the capability of increasing their contribution to the field. An expanding number of them are using Charge Coupled Devices (CCD) and can reach magnitude 20 and 21 - magnitudes only attainable with large professional telescopes until a decade ago. With this capability a new area in astronomy is being pioneered and is now opening the search for gamma ray burst (GRB) afterglows. The AAVSO has a set up GRB international network of amateur astronomers to search for GRB afterglows. Already four amateur astronomers have successfully imaged GRB afterglows.

One of the major challenges in modern-day variable star astronomy is not so much the discovery, but rather the tracking of new found objects. Each of these stars should be cataloged and named properly. However, we find that the cataloging and naming of the tens of thousands of new variable stars is not fully accomplished. The problem is partly due to a lack of funds, the enormoous scope of the problem and the lack of satisfactory solutions. As a result, many new variable stars are not named and catalogued in an homogeneous system. To help combat this problem Commission 27 on Variable Stars of the IAU has set up a working group to come up with possible solutions. The data collected from these surveys should be compiled and made accessible in a user-friendly database. Adequate funding is necessary for the longevity of the observing programs and surveys, and for the data they produce.

At a time when ground- and satellite-based surveys are discovering variable stars at alarming rate, we need another Cecilia Payne-Gaposchkin with her amazing insight and ability "in contemplating and turning over in [her] hands and in [her] mind" all the information available to create order from a tangled web of variable stars.

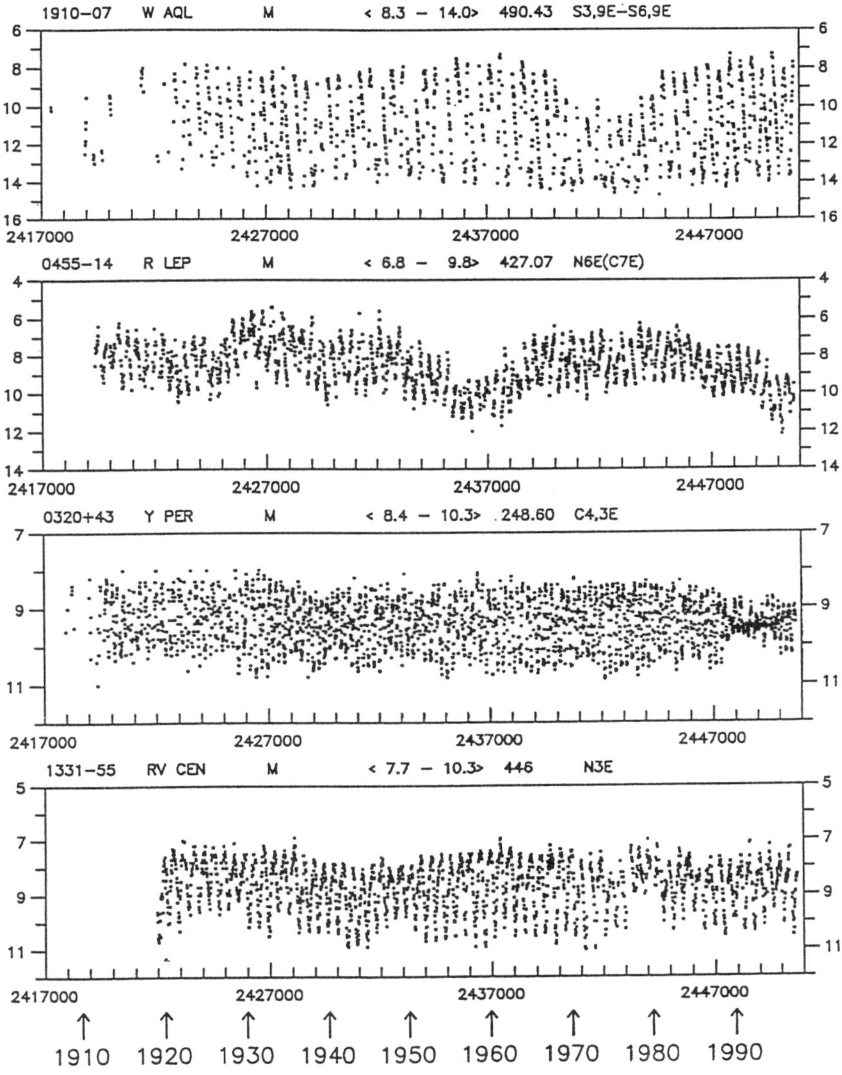

Fig. 2. Lightcurves for W Aql, R Lep, Y Per and RV Cen from the AAVSO International Database.

REFERENCES

ESA. 1997, The Hipparcos and Tycho Catalogues, Hipparcos Variability Annex: Light Curves, ESA SP-1200, Vol. 12

Hoffleit, D. 2000, Journal of the AAVSO 28, 2, 127

Kazarovets, A. V., Samus, N. N., Durlevich, O. V., Frolov, M. S., Antipin, S. V., Kireeva, N. N. & Pastukhova, E. N.1999 Information Bulletin of Variable Stars, No. 4659

Kazarovets, E. V., Samus, N. N., & Durlevich, O. V. 2000, Information Bulletin of Variable Stars, No. 4870

Kholopov, P. N. 1982 New Catalogue of Suspected Variable Stars, NAUKA, Moscow

Kholopov, P. N. et al. 1985 - 87, General Catalogue of Variable Stars, 4th ed., NAUKA, Moscow

Kidwell, P. 1996 in Cecilia Payne-Gaposchkin: An Autobigraphy and Other Recollections, K. Haramundanis, ed., 2nd ed., Cambridge University Press, Cambridge, p. 11

Payne-Gaposchkin, C. 1957 The Galactic Novae, North-Holland Publishing Co., Amsterdam

Payne-Gaposchkin, C. 1978 ARA&A 16, 1

Payne-Gaposchkin, C. 1996 in Cecilia Payne-Gaposchkin: An Autobiography and Other Recollections, K. Haramundanis, ed., 2nd ed., Cambridge University Press, Cambridge, p. 68

Payne-Gaposchkin, C. & Gaposchkin, S. 1938 Variable Stars, Lew A. Cummings Co., Manchester

Payne-Gaposchkin, C., Gaposchkin, S. & Menzel, D. 1952a Annals of Harvard College Observatory, 115, 1

Payne-Gaposchkin, C., Gaposchkin, S. & Menzel, D. 1952b Annals of Harvard College Observatory, 118, 1

Samus, N. N. 1990 General Catalogue of Variable Stars, Vol. 4, 4th ed., NAUKA, Moscow

EARLY GALACTIC NUCLEOSYNTHESIS: THE HALO FIELD STARS

Christopher Sneden

Department of Astronomy & McDonald Observatory, University of Texas, Austin, Texas

ABSTRACT: Recent abundance studies of field metal-poor halo stars shed new light on early galactic nucleosynthesis. This review brings attention to the changes in CNOLi elements with evolutionary state in low-mass stars of the metallicity range a tenth to a hundredth of solar, to the detailed neutron-capture abundance patterns of extremely metal-poor stars and to some recently discovered anomalies in the α-elements in several other halo stars.

1. INTRODUCTION

Cecilia Payne-Gaposchkin's Ph.D. thesis (Payne 1925) essentially created the branch of astrophysics devoted to stellar chemical compositions and the origin of the elements. She was the first to fully explore the application of the Boltzmann and Saha results to the spectra of stars and (especially) the Sun. Today many astronomers remember her thesis mainly for its bold assertion that, "The uniformity of composition of stellar atmospheres appears to be an established fact" (p. 189), a conclusion that subsequently proved to be correct for most stars. The thesis also noted (p. 188) that, "The outstanding discrepancies between the astrophysical and terrestrial abundances are displayed for hydrogen and helium", and guessed wrongly that, "The enormous abundance derived for these elements in the stellar atmosphere is almost certainly not real". But this statement undoubtedly was influenced by H. N. Russell, who was unconvinced that hydrogen could be so overwhelmingly abundant in the universe (e.g., see the discussion by Gingerich 1982). Casual study of Payne-Gaposchkin's thesis reveals reveals many other fascinating aspects worthy of comment. First and foremost, it must be emphasized that the thesis reads extremely well 75 years after its completion. It is written with clarity and economy, and even includes an "Index to Definitions" for specialized terms such as "marginal appearance" of spectral lines; all readers of the document should be grateful for this. Her thesis ought to be required reading for anyone beginning a scientific publishing career.

Second, the spectroscopic details discussed in Payne-Gaposchkin's thesis should remind us that in the early 1920's, large gaps existed in basic atomic and nuclear physics data. Here are a few examples.

The Starry Universe: The Cecilia Payne-Gaposchkin Centenary
A. G. Davis Philip & Rebecca A. Koopmann, eds.
p. 77-86

Christopher Sneden
University of Texas
Department of Astronomy
Austin, TX 78712
CHRIS@VERDI.AS.UTEXAS.EDU

Her Table I gives mass numbers of individual isotopes for only 50 of the 84 listed elements, terrestrial abundance percentages for only 32 of these elements and acknowledges stellar/solar spectrum detection of only 43 of them. Table III contains first ionization potentials for only 47 elements (a couple more are given in Table XVII). Blank entries in Table III include the ionization potential of neutral carbon, and she acknowledges on p. 59 that the spectrum of C I had not yet been classified in the laboratory, so that its presence in stellar spectra could not be sought. Rare earth element transitions had already been seen in stellar spectra, but on p. 86 she notes that, "No series relations have as yet been published for any rare earth element, excepting a short list of relative term values for lanthanum. From analogy with the previous long period it would seem unlikely that the first ionization potential of these atoms can be as great as 13 volts, ...". She was right: we now know that the ionization potentials of the rare earths lie in the range from 5.5 to 6.3 eV. But her statement illustrates how often her thesis results depended as much on educated guesses about atomic physics as on well-determined laboratory data. Given this situation, it is remarkable that Payne-Gaposchkin was able to bring coherence to solar/stellar spectroscopy and give it a clear physical basis that set the stage for future workers.

A brief review of current stellar chemical composition analyses cannot do justice to all the topics touched on by Payne-Gaposchkin's thesis. This paper will be limited to abundances of field stars of the galactic halo and their application to early galactic stellar nucleosynthesis. In succeeding sections we consider the light elements CNOLi, which are susceptible to proton-capture fusion, the neutron-capture elements ($Z > 30$) and some anomalies in the α-capture elements, all suggesting promising areas for future study.

2. CNOLi ELEMENTS: "MIXED" RESULTS

We first consider "mixed" results in the sense of the alterations in surface abundances from the convective mixing of light elements vulnerable to hydrogen burning fusion (the CNO cycle, etc.) in stellar interior regions that are hotter than $T \sim 10^6$ K. For solar metallicity stars in the mass range 1 - 10 $\mathcal{M}_\odot$, Iben (1967 and references therein) predicted the changes in surface abundances of these elements in the convective mixing "first dredge-up" episode during the subgiant and lower red giant branch evolution. Surface abundances of C were predicted to drop by about a factor of two and N abundances would increase in response; carbon isotope ratios would decrease from $^{12}C/^{13}C \sim 90$ to ~ 25; and O would be little changed. These predictions were confirmed quantitatively in a series of high resolution spectroscopy papers culminating in Lambert & Ries (1981) and Lambert et al. (1980). But it was noted in these papers (see also Cottrell & Sneden 1986, Shetrone et al. 1993) that low-mass, mildly metal-poor old disk giants had much lower isotopic ratios $^{12}C/^{13}C \sim 7$ - 15, and less altered C and N abundances. But galactic halo giants exhibit much more extreme H-burning abundance products at their surfaces: $^{12}C/^{13}C \sim 4$ - 8 (Sneden et al. 1996); C-to-N conversions of factors of ten or more (Kraft et al. 1982). This indicates that more extensive envelope processing through the H-burning layers has occurred in these low mass stars compared to higher mass disk stars.

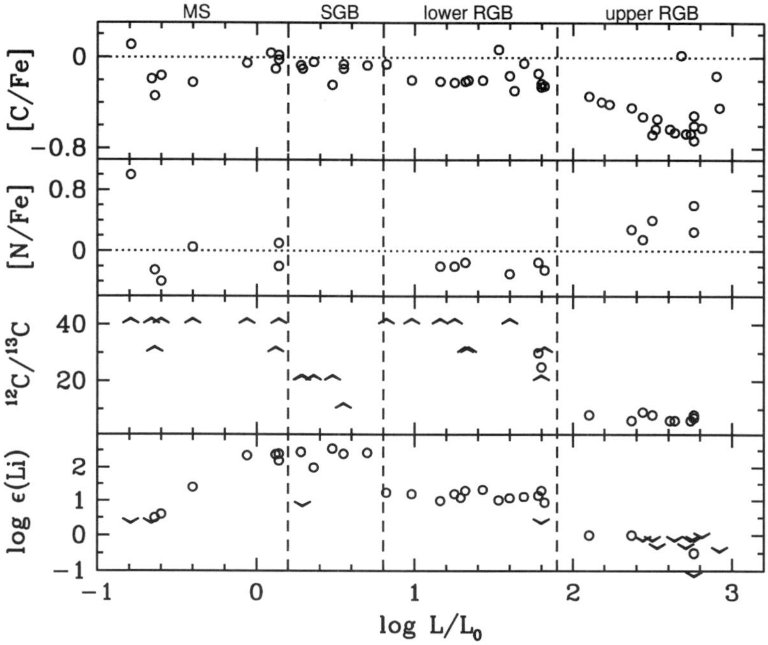

Fig. 1. Abundance trends of light elements with stellar luminosity. This figure is adapted from Gratton et al. (2000). Lower limits to $^{12}C/^{13}C$ ratios from non-detected spectral features of ^{13}C are depicted by upward-pointing heads of arrows, and upper limits to log ϵ(Li) from non-detected Li I lines are depicted with the same symbols pointing downward. The vertical dashed lines separate the different evolutionary domains that are noted at the figure top (MS for main-sequence stars, etc.).

From advances in high-resolution spectroscopy, and increased accuracy in luminosities of metal-poor field stars (due to Hipparcos parallaxes, Perryman 1997, and Strömgren photometry, Anthony-Twarog & Twarog 1998), a coherent picture now exists on the evolution of the light elements in these stars. This is illustrated in Fig. 1, adapted from Gratton et al. (2000). This figure shows C, N, $^{12}C/^{13}C$ and Li abundances as functions of luminosity (hence evolutionary state) for stars in the metallicity range -1 > [Fe/H] >–2. The dashed line at log $L/L_\odot$ = 0.2 marks the approximate luminosity of the Pop II main-sequence turnoff; the one at log $L/L_\odot$ = 0.8 shows luminosity of the giant branch base, where the first dredge-up effectively has been completed; and the one at log $L/L_\odot$ = 1.9 indicates the approximate position of the luminosity function "bump" in globular clusters.

For [C/Fe], [N/Fe] and $^{12}C/^{13}C$, it is clear that no significant surface abundance changes occur through the first-dredge-up and beyond, but large alterations in all three ratios can be detected suddenly near log $L/L_\odot$ = 1.9. The fragile element Li, susceptible to proton captures at relatively low temperatures, reveals more complex behavior. In lower main-sequence stars (log $L/L_\odot$ < 0) Li abundances are small probably because during the very short lifetimes of such stars, slow circulation currents cycle essentially all the envelope material down to Li fusion temperatures.[1] Near the main-sequence turnoff and subgiant evolutionary domains, the Li abundance is constant at the now-familiar "Spite plateau" value (for recent work on this important quantity for Big Bang nucleosynthesis, see e.g., Ryan et al. 2000 and references therein). Near log $L/L_{\odot}$ = 0.8 the Li abundance is seen to drop by a factor of about ten, and to then remain steady until much further evolution. The lack of first dredge-up effect on the surface abundances of anything but Li is in excellent accord with standard stellar evolution theory (e.g., Sneden et al. 1996), which predicts that low-mass, low-metallicity subgiants will have relatively shallow convective envelopes that are unable to dig deep enough to dredge the main CNO-cycle products to the stellar surfaces.

Further along the red giant branch (at log $L/L_{\odot} \sim 1.9$) another Li abundance decrease of at least a factor of ten accompanies the abrupt changes in [C/Fe], [N/Fe], and $^{12}C/^{13}C$. These changes appear to happen near the "bump" region in the globular cluster luminosity function. The explanation seems to be clear: the bump is thought to be caused by the small pause in the red giant evolution of low mass stars as their outward-advancing H-burning shells eat through the composition (hence mean molecular weight) discontinuity left behind in the envelope after completion of the first dredge-up. Apparently this act also is accompanied by or drives a new mixing episode that significantly alters the CNOLi abundances that exist after the first dredge up. The case is strong, but theoretical predictions of the bump mixing episode with self-consistent physical bases are currently lacking.

The second meaning of "mixed" results refers to the ongoing debate on the true level of O abundances in metal-poor stars. All spectroscopists who have analyzed the ground-state [O I] lines, the high excitation O I triplets, the OH electronic UV band lines and the OH vibration-rotation IR band lines are in agreement that [O/Fe] > 0.0 in almost all stars with [Fe/H] < -0.5. Frustratingly, the accord ends there, as depicted in Fig. 2. Analyses of the [O I] and OH IR lines consistently yield [O/Fe] $\sim$ +0.3 to +0.5, essentially independent of (or very weakly dependent on) [Fe/H], while results from the OH UV and O I triplet lines just as consistently yield [O/Fe] values that increase strongly with decreasing metallicity.[2] This is a serious concern: not only is the [O/Fe] ratio necessary to constrain models of massive-star early galactic nucleosynthesis, but O is an important opacity source in stellar envelopes, and so its abundance directly affects the photometrically derived ages of globular clusters. The O abundance problem was considered in Joint Discussion #8 at the IAU 2000 General Assembly, and the interested reader should consult the proceedings of JD #8 (Barbuy 2001) for the pros and

[1]An intriguing aspect of Payne-Gaposchkin's thesis lies in spotting all the near-discoveries of chemical composition effects whose implications were not able to be fully understood due to spectroscopic data or atomic physics limitations of that era. In the case of Li in the solar spectrum, she writes on p. 59, "Russell has called attention to the fact that [the 6707 Å Li I resonance doublet] is fainter, in the Sun, than would be anticipated from the terrestrial abundance of the element". She goes on to quote another investigator's incorrect explanation for the Li I line weakness involving Doppler effect blurring of this feature, but actually she was seeing evidence of main-sequence Li destruction. It is tempting to speculate, though impossible to prove, that Payne-Gaposchkin would have eventually hit upon the correct nucleosynthesis idea underlying this and other stellar spectroscopic anomalies, had her research attention not turned to variable stars.

[2]Payne-Gaposchkin's thesis does not directly address questions of stellar O abundances, as only the high excitation O I triplet at 7770 Å had been detected in the solar spectra available to her. She did correctly surmise that the triplet lines should behave in a roughly similar fashion to the H I Balmer lines, and thus should reach their maximum strength in A-type stars.

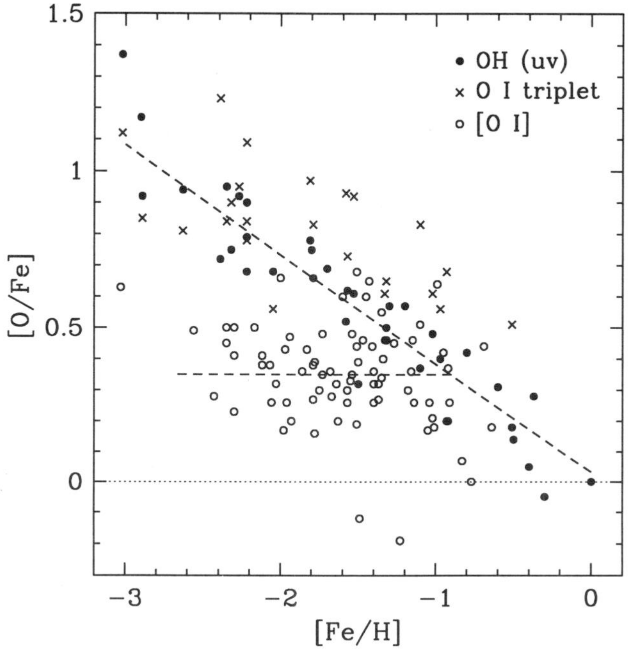

Fig. 2. [O/Fe] ratios as a function of [Fe/H] metallicity, as deduced from different O species. The results for OH and O I triplet features are taken from Boesgaard et al. (1999) and Israelian et al. (1998), while those from the [O I] lines are from Gratton et al. (2000). The slanting dashed line is a best-fit line to the OH results recommended by Boesgaard et al., and the horizontal dashed line represents the mean value from the [O I] lines.

cons of this issue. It is suggested here that the [O I] features will turn out to be the most reliable O abundance indicators, especially when referenced to abundances of Fe-peak metals deduced from ionized lines (e.g., Sneden & Primas 2001). When analyses are done in this fashion, the uncertainties in all atmospheric parameters (T_{eff}, log g, v_t) virtually cancel out, leaving [O/Fe-peak] abundance ratios dependent mostly on simple line strength ratios. This view is admittedly a personal bias, and a lot more work is needed (especially in understanding the formation of [O I], O I, OH UV and OH IR lines) to bring a coherent answer to the question of the O abundance in the early Galaxy.

3. NEUTRON-CAPTURE ABUNDANCE STUDIES

The elements with atomic numbers $Z > 30$ are often called "neutron-capture" (n-capture) elements because their most abundant isotopes are synthesized in neutron bombardment reactions. When the neutron flux is low enough that unstable nuclei can undergo β-decays between successive neutron captures, production of the elements proceeds up the valley of β-stability. This mechanism, called the $s[low]$-process, occurs during helium burning phases of the late evolutionary phases of low-intermediate mass stars. Alternately, in conditions of short-lived, extremely large neutron fluxes, the time scale for neutron captures is much smaller than that for β-decays, and ultra-neutron-rich nuclei are created out to the "neutron drip line", only to decay back quickly to the valley of β-stability when the neutron flux ceases. The site of this "*r[apid]*-process" has yet to be determined, but probably is associated with the extreme conditions associated with the deaths of massive stars.

The r- and s-processes can manufacture different amounts of the various isotopes of the n-capture elements, so the total elemental abundances resulting from these processes can also be quite different. Thus abundance studies of metal-poor stars that determine the relative contributions of these two processes to the heavy elements can be used to describe the influence of large versus low-intermediate mass stars to the overall nucleosynthesis in the early Galaxy. Of course, not all of the more than 50 n-capture elements can be detected in stellar spectra. Much of the work on these elements has centered on studies of the ionized species of the easily-detected rare earths. Recent studies have concentrated on either determining a few abundances in large samples of metal-poor stars (e.g., McWilliam et al. 1995, Ryan et al. 1995, Burris et al. 2000), or detailed abundances sets for a few stars (e.g., Sneden et al. 2000a, Westin et al. 2000).

Two clear conclusions emerge from the large-sample studies: the bulk amounts of the n-capture elements (i.e., $<$[rare-earths/Fe]$>$) vary by more than two orders of magnitude in stars with the same approximate metallicity, at least for stars with [Fe/H] < -2. First, the derived star-to-star neutron-element abundance scatter is far larger than observational errors, which suggests that the formation of very metal-poor stars occurred in an inhomogeneous early galactic halo ISM, with localized enrichment in n-capture elements from early massive-star nucleosynthesis. Second, the mean ratio $<$[Ba/Fe]$>$ exhibits a decline from $\simeq$0.0 at [Fe/H] ≥ -2 to $\simeq -0.8$ at [Fe/H] $\simeq -3$. Since Ba isotopes can most easily be made in the s-process while those of Eu are usually products of the r-process, the shift in [Ba/Eu] toward the low metallicity end of halo stars indicates the increasing dominance of r-process synthesis from massive, short-lived progenitor stars in the early Galaxy.

A few stars are so (relatively) rich in n-capture elements that abundances may be derived for a large number of these elements. Stars such as CS 22892-052 (Sneden et al. 2000a) and HD 115444 (Westin et al. 2000) are excellent spectroscopic laboratories of n-capture element transitions. Indeed, Payne-Gaposchkin's thesis noted that, "The spectra of the rare earths are so rich in lines that spurious identifications with lines in stellar spectra are often likely to occur". This makes abundance determinations in n-capture-rich stars often laborious, with synthetic spectrum computations needed to disentangle the components of sometimes quite blended n-capture transitions (e.g., Figs. 3 and 7 of Sneden et al. 1996). But the few abundance determinations of very metal-poor giants that include detailed analyses of $>$10 n-capture elements show that almost the entire abundance run for $Z \geq 56$ may be fit well with a scaled solar r-process-only abundance distribution.

A significant exception to the excellent fit of the heavier n-capture element distribution to the solar r-process curve is thorium, whose abundance is about a factor of two smaller compared to that predicted by the abundances of the rest of the $Z \geq 56$ elements. This is because Th radioactively decays with a half-life of 14 Gyr. Butcher (1987) first attempted to derive a galactic disk age from

Th abundances of relatively metal-rich stars, and François et al. (1993) applied the same idea to metal-poor halo stars. Turning an observed [Th/stable-n-capture] ratio into an "age" of the n-capture material for a given star is not easy, and depends on reliable predictions of the initial amount of Th created in the prior n-capture synthesis event (e.g., Cowan et al. 1999, Goriely & Clerbaux 1999). The best estimates for the two stars mentioned previously and for three giants of the globular cluster M 15 (Sneden et al. 2000b) lie in the 12 - 16 Gyr range. But the very recent discovery of another extremely n-capture-rich halo star with detection of both Th and U ($\tau_{1/2}$ = 4.5 Gyr for the longer-lived ^{238}U isotope) by Cayrel et al. (2001) could potentially settle the issue. The "absolute" production of either Th or U is difficult to predict accurately, but not so the production *ratio* of these two neighboring elements. Cayrel et al. derive an age of 12.5 $\pm$ 3.3 Gyr from a preliminary analysis of the [Th/U] ratio in their star. The uncertainty in this age can be made smaller from a more detailed investigation, and surely there are other very metal-poor stars with large n-capture contents; for this reason alone it is imperative to continue the large-sample spectroscopic surveys of the galactic halo.

4. SOME OTHER ABUNDANCE SIGNALS: NEW DIRECTIONS AND CURIOSITIES

This brief review of early galactic nucleosynthesis from the spectroscopy of halo field stars ends now in a fashion that we hope would have pleased Cecilia Payne-Gaposchkin: unfinished business. As indicated previously, her thesis contains reference to spectroscopic puzzles that she knew could not be immediately answered, but were surely worth mentioning for future workers. Here attention is called to another puzzle: the existence of low-α halo stars. Almost all stars more metal-poor than [Fe/H] $\sim$ −0.8 have enhanced abundances of those light elements whose dominant isotopes are simple multiples of α particles. These elements include the spectroscopically observable O, Mg, Si, Ca and Ti and they are almost always overabundant in halo field and cluster stars by a factor of two or more, e.g., [α-capture/Fe] $\simeq$ +0.35 $\pm$ 0.15. The α-capture elements are products of massive-star nucleosynthesis, and their overabundance is one more sign of the dominance of these short-lived element donors to the galactic halo ISM.

Recently however, exceptions to the high-α rule for halo stars have been discovered. Carney et al. (1997) found that BD +80° 245 has [Fe/H] $\simeq$ −2, but [Mg, Si, Ca, Ti/Fe] $\leq$ 0, and Nissen & Schuster (1997) derived the same low-α trend in more metal-rich halo stars ([Fe/H] $\sim$ −1) that possess orbits which take them to the outer halo regions of the Galaxy. Brown et al. (1997) argued that giants of the outer halo globular clusters Rup 106 and Pal 12 are similarly weak in α-capture elements. New evidence now suggests that the class of low-α halo stars is more complex than previously thought. In Fig. 3 we compare three low-α stars (BD +80° 245, Carney et al. 1997, G4-36, James 2000, CS 22966-043, Preston & Sneden 2000) whose abundance patterns are currently undergoing detailed scrutiny by our group. This figure contains our preliminary results, but the main points are already clear. First, the typical halo field stars of this metallicity (Cayrel 1996) have [Mg,Si,Ca,Ti/Fe] $\sim$ +0.3, whereas there are significant deficiencies of Mg, Si and Ca in all three low-α stars. Second, BD

+80° 245 also has [Ti/Fe] < 0.0 and "normal" Fe-peak abundances for a halo star, but G 4-36 and CS 22966-043 have overabundances of all these elements compared to Fe. Finally, all three low-α stars have extreme (~1 dex) deficiencies of the n-capture elements Sr and Ba; the abundances of other n-capture elements are too low to be detected.

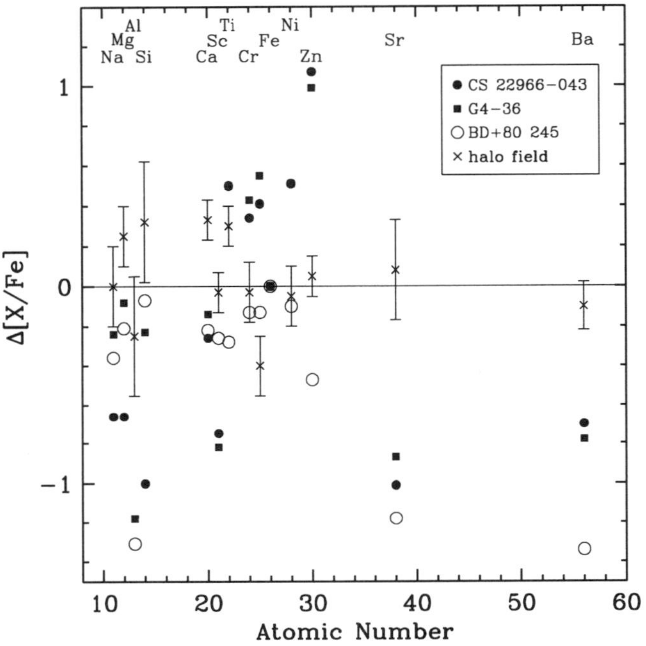

Fig. 3. Abundances of α-poor stars compared with typical halo field stars. All objects in this figure have [Fe/H] ~ -2, and the symbols representing them are given in the figure legend.

The nucleosynthetic origin of the low-α stars is unclear at present. Type I supernovae are thought to produce more Fe-peak elements than α- or n-capture elements, but attempts to match the abundances of these stars with predicted Type I nucleosynthesis yields (e.g., Iwamoto et al. 1999) are so far not promising. It is also not clear how such metal-poor material (presumably the low-α stars formed shortly after the Galaxy's birth) would have been able to be enriched by the long-lived low-mass stars that are thought to end as Type I SNe. Clearly more rigorous confrontation of theoretical models is needed, and with three low-α stars now known, these must represent a non-negligible part of early galactic nucleosynthesis; they cannot be ignored in overall galactic chemical evolution modeling.

ACKNOWLEDGMENTS

The work reviewed here has been a joint effort of many people. The contributions and inspiration of those cited in the reference list and other collaborators are very appreciated This research has been supported by the US National Science Foundation, most recently by grant AST-9987162.

REFERENCES

Anthony-Twarog, B. J. & Twarog, B. A. 1998 AJ 116, 1922

Barbuy, B., ed. 2001 Oxygen Abundances in Old Stars and Implications to Nucleosynthesis and Cosmology, JD #8 of IAU Gen. Assbly. XXIV, ASP Conf Ser. and New Astr. Rev., in press

Boesgaard, A. M., King, J. R., Deliyannis, C. P. & Vogt, S. S. 1999 AJ 117, 1999

Brown, J. A., Wallerstein, G. & Zucker, D. 1997 AJ 114, 180

Burris, D. L., Pilachowski, C. A., Armandroff, T. A., Sneden, C., Cowan, J. J. & Roe, H. 2000 ApJ 544, 302

Butcher, H. R. 1987 Nature 328, 127

Carney, B. W., Wright, J. S., Sneden, C., Laird, J. B., Aguilar, L. A. & Latham, D. W. 1997 AJ 114, 363

Cayrel, R. 1996 A&ApRv 7, 217

Cayrel, R. et al. 2001 Nature, in press

Cottrell, P. L. & Sneden, C. 1986 A&Ap 161, 314

Cowan, J. J., Pfeiffer, B., Kratz, K. -L., Thielemann, F. -K., Sneden, C., Burles, S., Tytler, D. & Beers, T. C. 1999 ApJ 521, 194

François, P., Spite, M. & Spite, F. 1993 A&Ap 274, 821

Gingerich, O. 1982 in Symposium on the Orion Nebula to Honor Henry Draper, A. E. Glassgold, P. J. Huggins & E. L. Schucking, eds., Ann. New York Acad. Sci., 395, 395

Goriely, S.& Clerbaux, B. 1999 A&Ap, 346, 798

Gratton, R. G., Sneden, C., Carretta, E., & Bragaglia, A. 2000 A&Ap 354, 169

Iben, I. Jr. 1967 ARA&Ap 5, 571

Israelian, G., García Lopez, R. G. & Rebolo, R. 1998 ApJ 507, 805

Iwamoto, K., Brachwitz, F., Nomoto, K., Kishimoto, N., Umeda, H., Hix, W. R., & Thielemann, F. -K. 1999 ApJS 125, 439

James, C. R. 2000 Ph.D. Disseratation, Univ. of Texas at Austin

Kraft, R. P., Suntzeff, N. B., Langer, G. E., Trefzger, C. F., Friel, E., Stone, R. P. S. & Carbon, D. F. 1982 PASP 94, 55

Lambert, D. L., Dominy, J. F. & Sivertsen, S. 1980 ApJ 235, 114

Lambert, D. L. & Ries, L. M. 1981, ApJ, 248, 228

McWilliam, A., Preston, G. W., Sneden, C. & Searle, L. 1995 AJ 109, 2757

Nissen, P. E. & Schuster, W. J. 1997 A&A 326, 751

Payne, C. H. 1925 Stellar Atmospheres (A Contribution to the Observational Study of High Temperature in the Reversing Layers of Stars), Harvard Coll. Obs. Mono. 1, ed. H. Shapley.

Perryman, M. A. C. ESA 1997 The Hipparcos and Tycho Catalogues (ESA Publ. Div., Noordwijk)

Preston, G. W. & Sneden, C. 2000 AJ 120, 1014

Ryan, S. G., Beers, T. C., Olive, K. A., Fields, B. D. & Norris, J. E. 2000 ApJ 530, 57

Ryan, S. G., Norris, J. E. & Beers, T. C. 1995 ApJ 471, 254

Shetrone, M. D., Sneden, C. & Pilachowski, C. A. 1993 PASP 105, 337

Sneden, C., Cowan, J. J., Ivans, I. I., Fuller, G., Burles, S., Beers, T. C. & Lawler, J. E. 2000a ApJ 533, L139

Sneden, C., Johnson, J., Kraft, R. P., Smith, G. H., Cowan, J. J. & Bolte, M. S. 2000b ApJ 536, L85

Sneden, C., McWilliam, A., Preston, G. W., Cowan, J. J., Burris, D. L. & Armosky, B. J. 1996 ApJ 467, 819

Sneden, C. & Primas, F. 2001 in Oxygen Abundances in Old Stars and Implications to Nucleosynthesis and Cosmology, JD#8 of IAU Gen. Assbly. XXIV, ASP Conf Ser., and New Astr. Rev., in press

Westin, J., Sneden, C., Gustafsson, B. & Cowan, J. J. 2000 ApJ 530, 783

REMINISCENCES OF CECILIA PAYNE-GAPOSCHKIN (1900-1979)

Dorrit Hoffleit

Yale University, New Haven, Connecticut

ABSTRACT: Cecilia H. Payne, one of the most gifted astrophysicists in the middle decades of the 20 century, was conspicuously discriminated against, in deference to not necessarily better males. Her greatest achievement, the determination that hydrogen is the most prevalent atom in the stars, was discredited because Henry Norris Russell had previously determined that the compositions of the stars were the same as that of the Earth's crust. To avoid conflicts, especially with Donald Menzel, Payne was diverted to working on stellar photometry. With her husband, she started the Milton Bureau for studying all variable stars brighter than 10 magnitude on Harvard patrol plates. She also analyzed several thousand variables in the Magellanic Clouds. In her work on variable stars she did not limit herself to photometry but discussed their spectra and probable evolution as well. In 1956 she was finally appointed Professor and made Chairman of the Department of Astronomy. A versatile person, she was also an expert in music and poetry.

1. INTRODUCTION

Cecilia Helena Payne was the most brilliant and at the same time the person most discriminated-against at Harvard College Observatory. She came to Harvard from Cambridge University, England, in 1923, after having been inspired by a lecture given in England by Director Harlow Shapley. Within only two years she gained the first Ph.D. degree in astronomy awarded at either Harvard or Radcliffe. A. Lawrence Lowell, President of Harvard University from 1909 to 1934 was adamantly against awarding any woman a Harvard degree. He felt a woman's place was in the home, not in academia. Radcliffe College had been founded to enable qualified women to get a Harvard education; all professors at Radcliffe were Harvard professors repeating their undergraduate courses for the women students. Hence Cecilia was awarded the degree from Radcliffe while having done all of the necessary work at Harvard Observatory.

Given an appointment at the observatory, Miss Payne had considerable influence on numerous graduate students. But the courses she taught could not be listed in the college catalogues under her name simply because she was a woman. Already in 1919 Dr. Alice Hamilton had been appointed As-

The Starry Universe: The Cecilia Payne-Gaposchkin Centenary
A. G. Davis Philip & Rebecca A. Koopmann, eds.
p. 87-97

Dorrit Hoffleit
Yale University
Department of Astronomy, PO Box 208101
New Haven, Connecticut 06520-8101
HOFFLEIT@ASTRO.YALE.EDU

sistant Professor in the Harvard Medical School, but this did not set a precedent for Harvard College. (Did the Medical School not come under President Lowell's jurisdiction?)

2. PREPONDERANCE OF HYDROGEN

Many years after the award of her degree in 1925, Otto Struve (1962), Director of the Yerkes Observatory, declared Cecilia's thesis the most brilliant ever written. Shapley had initially approved the thesis and then proudly showed it to Henry Norris Russell, recognized as the Dean of American Astronomers. He had been Shapley's mentor at Princeton, and Shapley seemed under the impression that Russell could never be wrong about anything in astronomy. Russell objected to Cecilia's conclusion that hydrogen was the most prevalent atom in the stars, as he himself had determined that the composition of the stars was closely the same as the composition of the crust of the Earth. Even Eddington (her former mentor in England) remarked that she must mean that hydrogen was the most common element on the stars [in the atmospheres], not in a star as a whole. She had to water down her conclusion by saying something must be wrong about her analysis because her conclusion was in contradiction to then current opinion [namely Russell's]. Ironically, in 1929 Russell himself, studying solar spectra at Mount Wilson, concluded that hydrogen is indeed the most profuse element in the Sun. He did remark that "The most important previous determination of the abundance of the elements by astrophysical means is that by Miss Payne", but he did not mention the fact that he had been responsible for her watering down her conclusions because he did not believe them. Some later authors erroneously credited Russell as being the first to establish this relationship for the Sun and stars. For example P. Moore (1987) wrote, "He [Russell] first suggested (correctly) that the Sun and stars are composed mainly of hydrogen". Actually Russell (1923) had found that the hydrogen lines were abnormally strong, but he had not deduced that hydrogen itself was remarkably abundant, but cited Saha as having attributed the strong lines to some sort of unexplained anomaly. In her autobiography Cecilia wrote, "If a piece of work received his imprimatur, it could be published; if not it must be set aside and its author had a hard row to hoe" (Haramundanis p. 177). In 1930 she wrote Russell asking for advice. She had been offered the post as head of the Physics Department at Barnard College, but despite discrimination at Harvard she was devoted to Shapley, and did not want to leave (Haramundanis 1984, p. 26, Rossiter 1982). For better or for worse, she did remain at Harvard the rest of her life.

3. CONFLICTS AND JEALOUSIES

A few weeks after Cecilia arrived at Harvard, Shapley informed her that one of Russell's students at Princeton, Donald Menzel, was already working on problems similar to hers. Russell had arranged for Donald Menzel to use the very same Harvard spectrum plates that Cecilia needed for her thesis. So Russell and Shapley agreed on what they considered a fair compromise, namely that Cecilia should work on early type spectra and Menzel on the cooler later types. This turned out to be unsatisfactory to both. In her autobiography (Haramundanis p. 162) Cecilia comments, "It was my first taste of jealousy, the struggle for priority. I had much to learn". Menzel earned his Ph.D. at Princeton in 1924 and worked at Lick Observatory from 1925 to 1932 when he came to Harvard as successor to Astrophysics Professor Harry Plaskett who was transferring to Oxford. Cecilia had deeply desired Plaskett's vacated position, declaring herself capable of teaching his courses. But as a woman, she was denied this post.

4. DIVERTED TO PHOTOMETRY

To avoid potentially more conflicts, Shapley eventually diverted Cecilia from astrophysics to photometry, an important field largely neglected after the decease of another pioneer, Henrietta Leavitt (1868 - 1921), whose final work was completed by Dr. Shapley's secretary, Arville Walker, in 1926 (Leavitt 1919 - 26). Leavitt had determined photographic magnitudes for sequences in all 50 Harvard Standard Regions, totalling 2042 stars, in most regions going fainter than 16.6 pg (Leavitt 1917). In 1931 Cecilia Payne derived photovisual magnitudes for all the Harvard Standard Regions except the North Polar Region and subsequently published numerous photometric papers on special regions, including derivations of red magnitudes and red color indices (Pane-Gaposchkin 1931 - 37), in addition to many photographic sequences derived for the Milton Bureau variable stars.

Between 1923 and 1942 Cecilia wrote (either alone or with co-authors) 78 papers dealing with the analysis of spectra; thereafter only for those dealing with the spectra of variable stars. From 1928 through 1942 she was author or co-author of 58 papers dealing with photometry (exclusive of spectrophotometry), of which 16 dealt with the elements of variable stars.

5. MY FIRST IMPRESSIONS

When I was first employed at Harvard Observatory in 1929 (to work on variable stars) my impression of Cecilia was that she appeared distant and seemed jealous of most of her colleagues. Indeed she experienced discrimination and jealousy from a very early age. She had a younger brother of whom she wrote in her autobiography (Haramundanis p. 86), "I felt from the first that my brother was the one who really mattered. My first reaction to my brilliant, warm-hearted brother Humfry was one of jealousy." One of my undergraduate classmates who worked for Cecilia, was actually afraid of her. Madalen told me that once, in a fit of temper, Miss Payne had picked up a pile of glass photographic plates and smashed them to the floor! That assistant soon resigned. It was incomprehensible to me how a person who was obviously highly respected could fail to have a happy disposition. Several years later, when Shapley had me look into reclassifying spectra of bright southern stars for the purpose of acquiring their spectroscopic absolute magnitudes, I felt she was even jealous of me, a person so very far beneath her in ability or accomplishment. Helen Sawyer Hogg, Radcliffe Ph.D., 1931, told me she, too, felt Payne's jealousy because Helen was writing her thesis under Dr. Shapley who was paying more attention to Helen than to herself, as he had done in her student days. It was only after her autobiography was published by her daughter, Katherine Gaposchkin Haramundanis (1984), that I finally really understood Cecilia (Hoffleit 1984). As a consequence of her experiences with discrimination, primarily because she was a woman, jealousy and ill temper would appear inevitable. Nevertheless, she was always very helpful to students and co-workers and I soon admired and respected her very much. That she became jealous of me must have been simply because I was using the very plates she was deterred from using, even though I used them for a relatively ulitarian purpose (determination of spectroscopic absolute magnitudes), in contrast to her highly advanced astrophysical needs.

6. HELPFUL TO GRADUATE STUDENTS

Many graduate students profited from her interest and advice on their theses and other publications. At least nine who earned the Ph.D. before 1940 acknowledged advice from Miss Payne or collaborated with her in published papers. The first Ph.D. in Astronomy awarded by Harvard was to Canadian Frank Hogg in 1929, who wrote his thesis on cometary spectra (Hoffleit 1999). Before leaving Harvard he collaborated with Payne on six published papers on stellar spectroscopy. Prior to 1940 four women after Payne received the Ph.D. from Radcliffe (Emma C. T. R. Williams 1930, Helen Sawyer 1931, Carol Anger 1932 and Dorrit Hoffleit 1938) and all acknowledged helpful counsel from Payne.

Shapley required graduate students to give popular lectures especially for school-age children, and sometimes public open night lectures for adults. On one such occasion early in my career, he suggested that I pay close attention to Mrs. Gaposchkin when she lectured. Her diction and organization were excellent, with the best vocabulary of anyone in the department. However, I noticed that she rarely talked directly to the audience. As I watched her, she never looked at the audience; she looked upward, as though she could see right through the ceiling to converse directly with the almighty above. I did not want to be that illusive of my listeners. But as for the organization of my talk, Shapley must have spoken to her, for she graciously invited me to dinner, after which I was to practice my first public lecture with her. She was generously helpful with suggestions.

In 1933 Miss Payne asked me to examine the available spectra of the unique variable, η Carinae, from which I could ascertain that the spectral changes were correlated with the photometric magnitudes (Hoffleit 1933). Then in 1937 she also asked me to check the literature and plates on the star τ Aurigae (Payne-Gaposchkin 1957), another peculiar star, similar to DQ Herculis, with characteristics of both novae and Algol type eclipsing binaries, later designated as variable types Nb+Ea. These were stimulating assignments.

7. ENTER SERGEI AND THE BEGINNING OF THE MILTON BUREAU

Dr. Payne gave a short communication on Forbidden Lines Observed in Novae at the Astronomische Gesellschaft in Göttingen, Germany, in 1933. There she met Sergei Gaposchkin who, as a persona non grata in Germany, pleaded with her to try to get him a job in the United States. Shapley at Harvard complied, and later in 1933 Sergei joined the staff. A few months later, April 1934, Cecilia and Sergei eloped. When she first proposed the systematic study of all known variables brighter than the 10th magnitude on the Harvard patrol plates (the origin of the Milton Bureau in 1934) I suspected it was largely on Sergei's behalf, although actually she had already been involved in both spectral analyses and photometric determinations of the elements of variables, having written or co-authored 15 papers on variables before 1934. The three quarto volumes of the Harvard Annals (Payne-Gaposchkin and Gaposchkin 1943-54) containing Milton Bureau results for some 1565 variables (to which over 30 part-time assistants contributed) still constitutes a gold mine for early epoch observations for those variables, especially for the ones with changing periods.

Wonderful as I feel the Milton Bureau output is (Hoffleit 2000), I cannot but think that someone with the same perseverance but not quite Cecilia's abilities could have done almost as good a job, whereas her outstanding, perhaps unique abilities in theoretical astrophysics went largely to waste in deference to a Russell male protégé. Russell in 1934 began thinking about retirement and whom he might train to become a competent successor by 1943. He lamented that the best candidate in America, alas, is a woman, meaning Cecilia (DeVorkin 2000). At that time, hiring a woman on either the Harvard or Princeton faculty was out of the question. Russell did not even mention Menzel as a possible candidate. Despite the fact that Russell was mainly responsible for Shapley's diverting Cecilia from theoretical astrophysics to photometry to avoid conflicts with Menzel, Cecilia was warm in her tributes to the memory of Russell at the time of the centenary of his birth (Payne-Gaposchkin 1977b and c).

8. ALAS, SMOKING

In her era little was said about the ill effects of smoking. Cecilia was the most prolific chain smoker I ever knew. At the end of every working day her 8-inch diameter ash tray was piled three inches deep with ashes and butts. Ultimately such smoking was presumably the greatest contributor to her final illness.

For what seemed like an interminable interval after my final Ph.D. oral examination by the astronomy faculty, I suffered intense fear that Cecilia's smoking might have cost me my degree. With a cigarette between her lips she asked a question, of which I could not grasp a single word. When I asked her to repeat the question, again, with the cigarette still tightly between her lips, all the response I got was so mumbled I still could not grasp what she was asking. I feared that the other faculty might have understood what she asked and therefore could have flunked me for not knowing the answer. But within half an hour of the completion of the examination, Dr. Shapley came to me and said "It's OK". I did not ask him if he had understood what her question had been.

Back in 1925 Cecilia, too, had an unhappy experience after her oral. Several hours after her examination, Miss Harwood found Cecilia weeping in her office. When Cecilia told Harwood she had not yet heard whether or not she passed, Miss Harwood went directly to check with Shapley. Cecilia was so brilliant nobody bothered to tell her she had passed, taking it for granted she would know!

9. MUSICAL TALENT

Cecilia' father, Edward J. Payne, had been a talented musician as well as an historian. Among all her other talents, Cecilia inherited his musical expertise. When Shapley discovered the 1879 manuscript by Winslow Upton, Observatory Pinafore, a parody on Gilbert and Sullivan's operetta, Pinafore, he planned a performance by the observatory staff and students for the next meeting of the American Astronomical Society in Cambridge in 1929. The part of the leading lady, Josephine, had to be performed by the most important lady astronomer at Harvard, namely Cecilia. She did a marvellous job, although I had a feeling it sounded a wee bit forced. In her autobiography poor Cecilia confessed, "Alas, it was a soprano part, and I was a contralto. No matter, it had to be done, with the result that I have never been able to sing since".

In 1949 Frances W. Wright had acquired a copy of Joseph Haydn's Toy Symphony (Wright 1987). She and two graduate students then conceived the idea of forming the OPO, the Observatory Philharmonic Orchestra. Chester Cook, a pianist, piano teacher, amateur telescope maker, and volunteer at the observatory, was asked to be the conductor. Frances bought a baby grand piano for the observatory library-lecture room, which she played whenever there were piano parts. After one enjoyable performance I asked Frances if the OPO would perform at one of the meetings of the Bond Astronomical Club of which I was then secretary. The Bond Club was an organization of armchair astronomers organized at the Observatory by Shapley in 1924 to help educate the public on astronomical advances.

Frances was happy to comply with my request (Wright p. 47). She had acquired some copies of music composed by Sir William Herschel, which would be performed for the Bond Club. Cecilia wanted to conduct this concert and Mr. Cook gladly took on the piano parts. The audience was highly appreciative. It was an all-Herschel performance, unique for this era, about 170 years after Herschel's discovery of Uranus in 1781, which marked the end of his professional musical performances in favor of full time astronomical endeavours.

10. PUBLICATIONS

Dr. Payne-Gaposchkin was author or co-author of 284 publications [1] of which eleven are treatises (Table 1). Of the articles, eight are with Menzel as co-author, written between 1933 and 1938, after

[1]Harumdanis (p. 239 - 255) lists 351 titles, but 67, giving data for individual variable stars, are included in just two articles in Harvard Bulletins (Payne 1928). The stars were listed separately in the Astronomischer Jahresbericht for 1928, one of the Haramdanis sources for the Payne-Gaposchkin Bibliography.

he became a member of the department at Harvard.

For a person as industrious and productive as Cecilia was in astronomy, one wonders where she found time to become as knowledgeable as she was on the literature of most languages, from ancient times forward. In the book on Variable Stars, compiled by both Gaposchkins, Cecilia headed each chapter with an appropriate quotation from poetry from England, America, Germany (Goethe), France (Voltaire), Italy (Dante), Russia (Pushkin) and from Latin and the Bible, all (except the Bible) in their original languages. Her text, "Introduction to Astronomy" (1954), similarly gave verses at the heads of most chapters, as well as some within the text. Here, however, all were translated into English as the book was intended for students who would probably not be versed in all the languages. The 1970 version of this text, revised by both Cecilia and her daughter, Katherine Gaposchkin Haramundanis, is considerably enlarged in view of many advances in research in the intervening years, especially for stars and galaxies. This volume has 23 chapters in contrast to 18 in the earlier edition. It has been said that her use of literary quotations was unique with Cecilia. However, Denison Olmsted (1791 - 1859) at Yale, likewise headed his chapters of Letters on Astronomy Addressed to a Lady (1841) with apt poetic quotations. None of his are the same quotations used by Payne-Gaposchkin, although they both had more than one quotation each from Alexander Pope, John Milton and Virgil. As a child Cecilia had practiced extensively writing verse; as an adult she did so only occasionally.

Cecilia in 1977 was the 29th speaker but the first woman to have been selected to deliver the prestigious Henry Norris Russell Lecture of the American Astronomical Society. She spoke on Fifty Years of Novae, commenting that she could not give any new results, but could offer only a panorama. Her lecture is published in the Astronomical Journal (1977a) and is highly recommended to anyone interested in novae. Written in her beautiful English style, it brings out the excitement not only of observing unexpected nova outbursts, but also in trying to interpret the observations physically. Let me quote from her two final paragraphs which again bring out her appreciation of past poets:

"The art of nova portraiture has recently been carried to the extreme by the astronomical counterparts of Pygmalion (an Ovidian, not a Shavian Pygmalion). The image of Galatea has been wrought of the most refractory and rebellious of materials. Is she coming to life before our eyes? Will the model not only resemble, but be the original? We are on the verge of knowing; the artists are now applying the final touches."

Many problems about novae were still not resolved, she concluded,

"So if you press forward with the observations, the next 50 years of novae will be at least as full of surprises as the last."

The supernova 1987A in the Large Magellanic Cloud (Genet et al. 1988) fulfills one such surprise, called the birth of neutrino astronomy.

One extensive book review Cecilia wrote with a feeling of great indignation was of "Worlds in

Collision" by Immanuel Velikovsky (1950). This is the longest book review ever published in Popular Astronomy. The editor, Curvin H. Gingrich, noted that the book had been favorably reviewed in a number of popular magazines, whereas the book comes into sharpest conflict with science, especially astronomy (Payne-Gaposchkin 1950, p. 278). Mythology is regarded by its author as proof of his mistaken theories. Cecilia herself stated, "On the face of it, the time spent [hers] might seem to be wasted, but the scholar has a duty towards the right advancement of learning". She shows that "The astronomical conclusions reached by Dr. Velikovsky are untenable and absurd" (Payne-Gaposchkin 1950).

Her final book, "Stars and Clusters", came off the press soon before her decease (Payne-Gaposchkin 1979, Hoffleit 1979). This brings the history of the study of variable stars up to that date, but it continues to remain excellent reading for scholar and amateur alike. Her literary bent is exemplified by the titles of the chapters, for example, 1. All the World's a Stage - the Galaxy; 5. Creeping like a Snail - The Main Sequence; 7. The Bubble Reputation - Supergiants; and 16. Finale - Pas de Deux. This comprehensive, easily readable volume demonstrates many of Cecilia's talents and the fact that although she was transferred from her first love, theoretical astrophysics, to photometry, she still pursued astrophysics and stellar evolution to the very end. Early hardships made her resourceful.

11. PROFESSORSHIP AT LAST

In her autobiography Cecilia comments that after Donald Menzel was appointed Director they finally became friends. In 1956 she was appointed full Professor as well as Chairman of the Department. (She confessed, however, that the chairmanship left her too little time for research. That could possibly be one reason why she was awarded the honor, if her predecessor had felt likewise.) In retrospect, she felt they should have been collaborators, not enemies, all those previous years. A significant reason why they could at last become friends may have been that now their fields of primary research no longer overlapped: he was working mainly on solar research, while she was deeply immersed in the properties of variable stars. To me, before I resigned in 1956 from my tenured appointment at Harvard, Menzel had said that variables were for amateurs, not professionals. He would not have dared to say that to Cecilia whose work, while it included what amateurs could do, dwelt additionally on the astrophysical properties, ages and evolutionary tracks of the variables (e.g., Payne-Gaposchkin 1974). After the Milton Bureau work was completed in 1954, she became author or co-author of over 25 additional articles on variable stars, including analyses of between 3000 and 4000 variables in the two Magellanic Clouds (Haramundanis p. 215). Her interest in observing the variables was not just to determine their types of variation and periods. Those results became the input data for probing the origin and evolution of the stars. In the final chapter of her last book she states "From a consideration of the dynamic evolution of globular clusters, it has been concluded that within 10^{10} years their critical cores must inevitably collapse, and massive supernovae will probably result... Such an event, a massive supernova, would be more spectacular than anything we have yet witnessed on the restricted stage of our galaxy. We await with anticipation the fulfillment of this prophesy."

Hers was an exceptionally scholarly and scientifically productive life. Much of her output continues to be valuable not only as part of the history of astronomy but as input data for continued modern investigations. But one wonders if she would not have accomplished additional more difficult astrophysical tasks if she had not been diverted into photometry in deference to male competitors in theoretical astrophysics.

12. HER SUCCESSFUL CHILDREN

Cecilia was devoted to her children, two boys and a girl. But she did not allow them to interfere with her career. The children, when they were not at school, were brought to the observatory where Shapley had no objection to their roaming about, playing hide and seek or whatever possessed them. Cecilia's office was on the top floor of Building D, where Shapley and some of the other high ranking professors had their offices. The children did not invade this area; but their father had an office on the ground floor of Building C, to which there was access not only through indoor passageways but also in or out of the window. If the children needed anything their father was always available. But at home their mother gave her children prime attention. The two older, Edward and Katherine, at some time became assistants at the observatory. Both boys earned Ph.D. degrees: Edward Michael is a geophysicist, the younger, Peter, is an astronomer now employed as a computer scientist. Katherine played an important role in the production of the massive four volume Smithsonian Astrophysical Observatory Star Catalog, giving positions and proper motions of 258,997 stars to 10th photographic magnitude (Whipple 1966).

13. ACKNOWLEDGMENTS

In conclusion I must express the gratitude that all of us must feel toward Katherine Gaposchkin Haramundanis for publishing her mother's autobiography together with her own reminiscences and tributes from a few of Cecilia's scholar friends and admirers: Jesse Greenstein, Peggy Kidwell and Virginia Trimble. The book, "Cecilia Payne-Gaposchkin, An Autobiography and Other Recollections", clarifies much that some of Cecilia's younger contemporaries might not previously have been able to understand.

I also wish to thank Owen Gingerich not only for inviting me to participate in this memorial, but also for alerting me to the fact that David DeVorkin's treatise on Henry Norris Russell has already been published. It confirms and enlarges upon my impressions of Cecilia's early troubles at Harvard.

14. BOOKS BY CECILIA-PAYNE GAPOSCHKIN

1925 Stellar Atmospheres, Harvard Monograph No. 1, Cambridge, England

1930 The Stars of High Luminosity, Harvard Monograph No. 3, McGraw-Hill, N.Y.

1938 Variable Stars (with S. Gaposchkin), Harvard Monograph No. 5, Cambridge, MA

1953 Stars in the Making, Eyre and Spottiswoode, London

1954 Variable Stars and Galactic Structure, Athelone Press, London

1954 Introduction to Astronomy, Prentice-Hall, New York

1957 The Galactic Novae, North Holland Publishing Co., Amsterdam

1970 Introduction to Astronomy (2nd. revised Ed., with K. Haramundanis), Prentice-Hall, Englewood Cliffs, New Jersey

1979 Stars and Clusters, Harvard Univ. Press, Cambridge, MA

1979 The Garnett Letters, privately printed [Payne's ancestors]

1979 The Dyers Hand, privately printed. Later incorporated into K. Haramundanis, Ed., "Cecilia Payne-Gaposchkin, an Autobiography and Other Recollections", 1st ed. 1984, 2nd.ed. 1999, Cambridge University Press

REFERENCES

DeVorkin, D. 2000 Henry Norris Russell, Dean of American Astronomers, Princeton University Press, p. 341

Genet, R. M., Hayes, D. S., Hall, D. S. & Genet, D. R. 1988 Supernova 1987A; Astronomy's Explosive Enigma, Fairborn Press. Mesa AZ

Haramundanis, K., ed. 1984 2nd Ed. 1999 Cecilia Payne-Gaposchkin, An autobiography and other recollections, Cambridge University Press

Hoffleit, D. 1933 Note on the Spectrum of Eta Carinae, Bull. Harvard Coll. Obs. No 893, 11

Hoffleit, D. 1979 The End of an Era: Cecilia Payne-Gaposchkin and Her Last Book [Stars and Clusters], Book Review, Journ. Amer. Assoc. Variable Star Observers, 8, 48

Hoffleit, D. 1984 Book Review of Haramundanis, K., ed., 1st ed., 1984, Cecilia Payne-Gaposchkin, An Autobiography and Other Recollections, Sky and Telescope, 68, 225

Hoffleit, D. 1999 Canadian Astronomers Who Earned the Ph.D. at Harvard in the Shapley Era, Jour. Ast. Soc. Canada, 93, 262

Hoffleit, D. 2000 The Milton Bureau Revisited, Journ. Amer. Assoc. Variable Star Observers 28, 127

Leavitt, H. 1917 The North Polar Sequence, Ann. Harvard Coll. Obs., 71, No. 3, and Harvard Standard Regions, Ibid, No. 4.

Leavitt, H. 1919 - 26, Standards of Magnitude for the Astrographic Catalogue, Ann. Harvard Coll. Obs. 85 Nos. 1, 7 and 8

Moore, P. 1987 The International Encyclopedia of Astronomy, Orion Books, N.Y., p. 367

Olmsted, D. 1841 Letters on Astronomy Addressed to a Lady, Marsh, Capen, Lyon and Webb, Boston

Payne, C. H. 1928a Elements for Fifty Long Period Variables, Bull. No 860, Harvard Coll. Obs., 1

Payne, C. H. 1928b Elements for Forty-two Long Period Variables, idem, No. 861, 8

Payne-Gaposchkin, C. H. 1931 - 37, Photometric Investigations, Ann. Harvard Coll. Obs. 89, Nos. 1 - 8 and 12

Payne-Gaposchkin, C. H. 1950 Worlds in Collision, Pop. Ast., 58, 278

Payne-Gaposchkin, C. H. 1954 Introduction to Astronomy, Prentice-Hall, N. Y.

Payne-Gaposchkin, C. H. 1957 The Galactic Novae, North Holland Publishing Co., p. 95

Payne-Gaposchkin, C. H. 1974 Distribution and Ages of Magellanic Cepheids, Smithsonian Contributions to Astrophysics, No. 16, 22

Payne-Gaposchkin, C. H. 1977a Fifty Years of Novae, Henry Norris Russell Prize Lecture of the American Astronomical Society, AJ 82, 665

Payne-Gaposchkin, C. H. 1977b Russell and the Composition of Stellar Atmospheres, in In Memory of Henry Norris Russell, A. G. D. Philip & D. DeVorkin, eds., Dudley Obs. Report No. 13, p. 15

Payne-Gaposchkin, C. H. 1977c After Dinner Address, in In Memory of Henry Norris Russell, A. G. D. Philip & D. DeVorkin, eds., L. Davis Press, Schenectady, p. 89

Payne-Gaposchkin, C. H. 1979 Stars and Clusters, Harvard Univ. Press

Payne-Gaposchkin, C. H. & Gaposchkin, S., 1943 - 54, The Milton Bureau Results, Ann. Harvard Coll. Obs. Vols. 113, 115 and 118

Payne-Gaposchkin, C. H. & Haramundanis, K. 1970 Introduction to Astronomy, Prentice-Hall, Englewood Cliffs, New Jersey

Rossiter, M. W. 1982 Women Scientists in America, Johns Hopkins Univ. Press, Baltimore and London, 210

Russell, H. N. 1923 Ionization and Pressure in the Reversing Layer of the Stars, Pop. Ast., 31, 22

Russell, H. N. 1929 On the Composition of the Sun's Atmosphere, Ap. J. 70, 11

Struve, O., & Zebergs, V. 1962 Astronomy of the 20th Century, Macmillan Co. 220

Velikovsky, I. 1950 Worlds in Collision, Macmillan Co.

Whipple, F. L. 1966 Smithsonian Astrophysical Observatory Star Catalog, Smithsonian Institution Washington, D.C., Vol. 4

Wright, F. W. 1987 Constant Vigilence, privately printed from an oral history, 45

STAR FORMATION IN THE NEARBY UNIVERSE

Margaret J. Geller

Smithsonian Astrophysical Observatory, Cambridge, Massachusetts

Elizabeth J. Barton

Herzberg Institute of Astrophysics, National Research Council of Canada, Victoria, B. C., Canada

Barbara J. Carter

Harvard-Smithsonian Center for Astrophysics, Cambridge, Massachusetts

ABSTRACT: Large redshift surveys provide a route to exploration of the large-scale processes which drive star formation in the nearby universe. We examine the role of tidally triggered star formation in a large sample of tight pairs of galaxies selected from a redshift survey.

1. DEDICATION

In 1952, a woman wrote "The *man* who presumes to talk of the evolution of the stars must needs to be an optimist with a sense of humor. And astronomers are incorrigible optimists. They peer up through a turbulent ocean of atmosphere at the stars and galaxies, forever inaccessible. They speak of million degree temperatures, of densities smaller than our lowest vacuum; they study light that left its source two hundred million years ago. From a fleeting glimpse, they reconstruct a whole history." The italics are ours and the woman was Cecilia Payne-Gaposchkin.

Cecilia Payne-Gaposchkin's book, "Stars in the Making", reaches from the Sun to the Coma cluster of galaxies. It discusses the evolution of stars and of galaxies. The book, written for a popular audience, is as profound as it is exquisitely written. It contains brilliant syntheses along with the misperceptions of its time. Its greatness, like the greatness of the woman who wrote it, is in the clear forging of links between star formation, stellar evolution and galaxy formation and evolution; these elements remain the foundation for our still fumbling attempts to understand the star formation

The Starry Universe: The Cecilia Payne-Gaposchkin Centenary
A. G. Davis Philip & Rebecca A. Koopmann, eds.
p. 99-108

Margaret J. Geller
Smithsonian Astrophysical Observatory
60 Garden St.
Cambridge, Massachusetts 02138
MGELLER@CFA.HARVARD.EDU

2. INTRODUCTION

In large part the story of galaxy formation and evolution is the story of the formation of stars from the ambient gaseous medium and the subsequent recycling of gas in the process of stellar evolution. Although HST and large telescopes on the ground have prompted a rush to study star formation in the universe at high redshift, the low redshift universe is a rich laboratory for examining the physical mechanisms of star formation.

From the perspective of large-scale structure research, one of the enticing aspects of the star formation problem is the importance of physical processes operating on vastly different spatial scales. For example, the x-ray emitting gas in clusters and groups of galaxies affects the star-forming properties of the galaxies which pass through it and the galaxies in turn affect their gaseous environment. The problem is one of great complexity. Galaxy-galaxy interactions may trigger star formation; passage through the intergalactic medium may quench it. In dense regions like the cores of rich clusters or even their outskirts, it is difficult to separate the effects of initial conditions, interactions with the intergalactic medium, and galaxy-galaxy interactions. All of these are undoubtedly important in determining the morphological mix of galaxies and the evolution of their stellar content over the history of the universe.

Here we consider some elements of the star formation puzzle. In Section 3 we discuss a complete spectroscopic survey of a regions of the nearby universe. As others have shown (Dressler 1980, Postman & Geller 1984), galaxies no longer forming stars dominate in dense regions of the universe like the cores of clusters of galaxies. A surprising and somewhat controversial result is that for galaxies currently forming stars, the EW(Hα), a measure of the strength and age of the young stellar population in a galaxy, is remarkably insensitive to the density of the environment (averaged over a scale of $\sim 5h^{-1}$ Mpc, the galaxy correlation length). This averaging scale is large compared with the scales of $\lesssim 100$ kpc of importance for the tidal interactions among galaxies which can trigger star formation.

To explore the impact of tidal interactions in tight pairs of galaxies on star formation, Section 4 discusses a spectroscopic survey of a large sample of tight pairs of galaxies selected from a redshift survey similar to the one described in Section 3. These data support a simple picture where a close pass initiates a burst of star formation. The burst ages as the pair separation increases. In Section 5 we connect the results of Sections 3 and 4. We conclude in Section 6.

3. STAR FORMATION IN A REDSHIFT SURVEY

Enormous improvements in instrumentation have enabled accumulation of redshift survey spectra which are uniformly adequate for spectroscopic classification of the galaxies in the survey. These data enable a first assessment of the link between galaxy environment and spectroscopic type and thus between galaxy environment and star formation.

The 15R survey covers the region $8^h \leq \alpha \leq 17^h$ in the two declination ranges $26.5° \leq \delta \leq 29.5°$ and $30.0° \leq \delta \leq 32.5°$. The survey is complete to $m_R \simeq 15.4$ and includes 3149 galaxies with uniform, high quality spectra (Carter et al. 2001, Geller et al. 2002). Fig. 1 shows a montage of the spectra in the survey ranked by EW(Hα). Vigorously star-forming galaxies occupy the lower portion of the plot. The trace below the plot shows the spectrum characteristic of these galaxies; the equivalent width of the Hα line is a luminosity-weighted measure of the fraction of young hot stars in the galaxy. Galaxies no longer forming stars occupy the upper portion of the plot. As the trace above the plot shows, absorption lines dominate their spectra. A fraction of the emission line galaxies are objects with active nuclei (AGN). Before examining the large-scale distribution of star-forming galaxies, we separate out the AGN using a technique first explored by Veilleux and Osterbrock (1987). This technique depends on evaluation of the line ratios [OIII]5007/Hβ and [NII]6583/Hα.

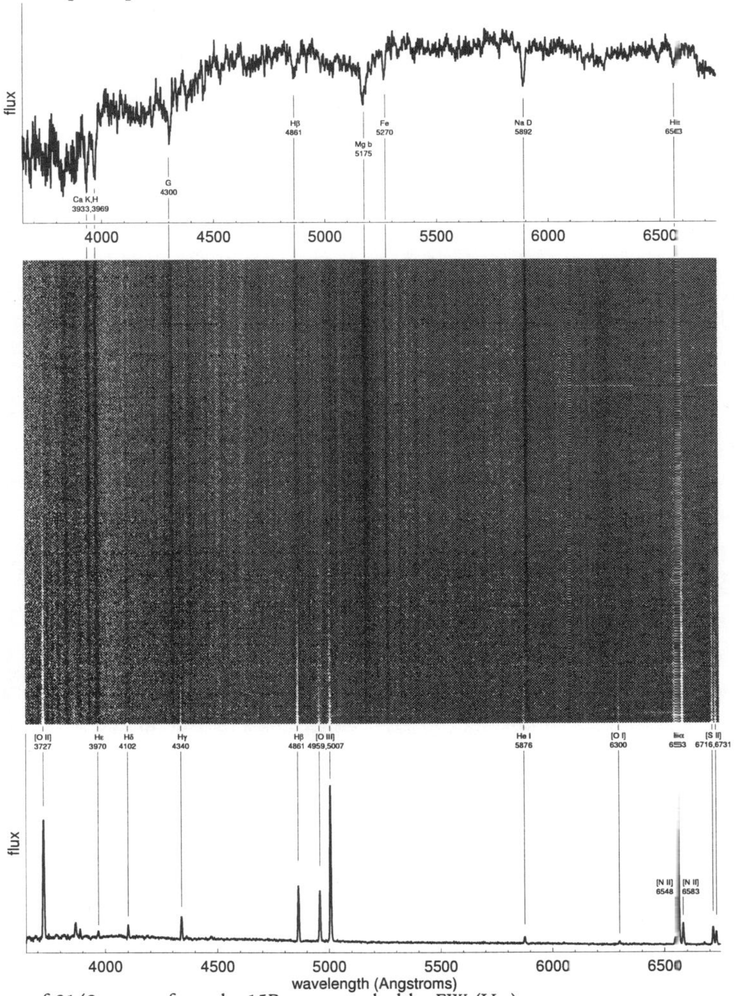

Fig. 1. Montage of 3149 spectra from the 15R survey ranked by EW (Hα)

Fig. 2 displays the distributions in redshift space of galaxies no longer forming stars (upper panel) and of star-forming galaxies (lower panel). The voids and thin walls characteristic of large-scale structure in the galaxy distribution are apparent in both panels. In the redshift space of Fig. 2, clusters of galaxies appear as "fingers" elongated along the line-of-sight (the radial direction) pointing toward the observer at the vertex of the plot. It is clear that star-forming galaxies are essentially absent from the most impressive of these clusters.

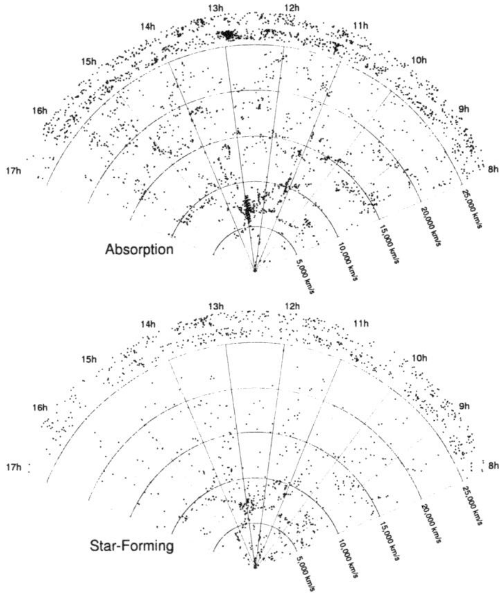

Fig. 2. Cone diagrams for absorption-line galaxies (upper panel) and star-forming galaxies (lower panel) in the 15R survey.

This result is another way of stating the well-known morphology-density relation: early-type galaxies are relatively more abundant in dense regions (Dressler 1980, Postman and Geller 1984). Morphology and spectroscopic type are correlated, but they are not the same. Segregation by spectroscopic type is a somewhat cleaner process than segregation by morphological type and a clear understanding of the relationship between the two awaits large catalogs which have uniformly and objectively defined morphological types (e.g., the Sloan Digital Sky Survey; York et al. 2000).

We can begin to investigate the relationship between environment and the star-forming properties of galaxies more deeply by asking whether the EW(Hα) depends on density. We estimate the density around each galaxy by computing the volume which contains its nine nearest neighbors and weighting appropriately by a selection function which accounts for our magnitude limited redshift survey. In essence, we compute the density smoothed over several Mpc. Fig. 3 shows a surprising result; for galaxies currently forming stars (they have EW(Hα) $\gtrsim$ 2 Å), the EW(Hα) is essentially independent of the surrounding smoothed density; there is only a slight tendency for increasing EW(Hα) with decreasing density.

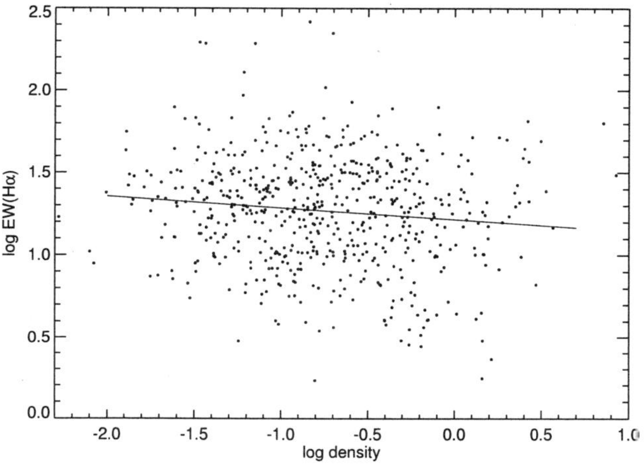

Fig. 3. EW(Hα) as a function of surrounding galaxy density.

The minimal trend in Fig. 3 is consistent with a similar analysis of the Stromlo-APM survey (Loveday, Tresse & Maddox 1999) and with the results of a survey of galaxies in nearby low density regions of the universe (Grogin & Geller 2000). It is apparently inconsistent with the much stronger correlation in the same sense in the LCRS (Hashimoto et al. 1998). The reason for the disagreement is unclear although it is interesting to note that both the 15R and Stromlo-APM surveys were done with long slits and the LCRS survey was done with fibers. The LCRS is more strongly affected by aperture bias (Kochanek, Pahre & Falco 2001).

The subtle dependence of EW(Hα) on the density (smoothed over a several Mpc scale) indicates that at least some environmental triggers of star formation may occur on a smaller, more "local" scale. The idea that tidal interactions between close pairs of galaxies triggers star formation is an old one (Toomre & Toomre 1972, Larson & Tinsley 1978), but now datasets are large enough and of high enough quality that we can begin to delve into the physics of the interaction. Next we examine the evidence for tidally triggered star formation in a large sample of tight pairs of galaxies selected from a redshift survey.

4. TIDALLY TRIGGERED STAR FORMATION

In their classic work, Larson & Tinsley (1978) compared a sample of normal galaxies with a set of interacting and peculiar galaxies to argue that "bursts of star formation are often caused by tidal interactions or collisions between galaxies". The data they used were limited to (U-B) and (B-V) colors. Now a host of photometric and spectroscopic studies at wavelengths from the radio to the x-ray examine sample of galaxies which often display some evidence of tidal distortion (e.g., Kennicutt et al. 1987, Sekiguchi & Wolstencroft 1992, Keel 1996, Donzelli & Pastoriza 1997).

Theoretical understanding of galaxy-galaxy interactions has advanced considerably since the first demonstration by Toomre & Toomre (1972) that tidal tails are resonance effects produced in prograde encounters. These simulations were n-body only. Mihos & Hernquist (1996, MH96) used n-body/SPH simulations to demonstrate the connection between nuclear bursts of star formation and tidal encounters. In encounters between two bulgeless galaxies of comparable mass, gravitational torques drive gas toward the center and initiate a starburst $\sim$ 100 Myr after the first close pass. In this case, bursts occur when the two galaxies are still separate systems. For galaxies with bulges, the starburst occurs only during the final stages of the merger. In these models, as in those of Toomre & Toomre, tidal tails appear in prograde encounters and are largely absent in retrograde encounters. For galaxies with shallow central potentials, a central burst of star formation occurs in *both* prograde and retrograde encounters. Dubinski, Mihos & Hernquist (1996) emphasize the sensitivity of easily observed tidal distortions to the physical parameters which describe the interacting galaxies and their orbits.

Large redshift surveys (e.g., Fig. 2) enable the identification of dynamically related tight pairs independent of the detailed morphology of the galaxies. In this spirit, Barton, Geller & Kenyon (2000, BGK) extracted a large, complete sample of tight pairs and n-tuples of galaxies from the CfA2 redshift survey which covers a larger solid angle to a shallower depth than the 15R survey (Geller & Huchra 1989, Huchra et al. 1990, Huchra, Geller & Corwin 1995, Falco et al. 1999).

The BGK pairs and n-tuples have velocity separations $\Delta V \leq 1000$ km s^{-1} and projected separations $\Delta D \leq 50\mathrm{h}^{-1}$ kpc. To isolate the effects of local interactions on star formation in these pairs, BGK focussed on pairs in regions with a smoothed density (analogous to the densities discussed in Section 3) less than 2.2 times the mean for the entire survey. These pairs in low density regions contain 162 galaxies.

High signal-to noise spectroscopy of the galaxies in tight pairs enables construction of a physical model for the connection between the dynamics of the pair and star formation in its member galaxies. For example, the EW(Hα) is a measure of both the age and the relative strength of a burst of star formation. For strong bursts with extended, constant star formation rates (constituting $\gtrsim$ 20% of the R-band flux), the EW(Hα) provides a robust estimate of the age of the burst, t_{burst}, which is insensitive to the presence of the old population in the galaxy. Thus, by examining the dependence of EW(Hα) on the kinematic parameters of the pairs, BGK establish a connection between the burst age and the tidal encounter.

Fig. 4 shows the fundamental observational signature of the relationship between burst age and pair dynamics. The EW(Hα) declines as both the velocity and projected spatial separation of the pair increase. These plots show only galaxies in sparsely populated environments. BGK observed all galaxies in pairs whether they have emission lines or not; absorption-line galaxies which no longer contain the ISM necessary for a burst of star formation populate the EW(Hα) = 0 line. Galaxies with relatively little gas and/or those which have had relatively ineffective encounters populate the region between the 0 baseline and the limiting curve. Pairs which have not yet had an encounter and interlopers also contribute to the zero baseline and the region below the limiting curve.

BGK use the models of MH96 to interpret Fig. 4. A close pass, a tidal encounter, initiates a burst of star formation. The burst ages as the galaxies move apart. The decrease in EW(Hα) associated with the aging of the burst accounts for the correlation BGK observe.

Fig. 4. EW (Hα) as a function of the projected pair separation, Δ D (left), and line-of-sight velocity difference, Δ V (right).

BGK use the starburst models of Leitherer (1999) to derive an upper limit on t_{burst} for each galaxy in the pair sample. They show that, as expected in the MH96 picture, the burst ages match the times of closest approach estimated from the projected separation and line-of-sight velocity difference.

The tie between burst age and dynamical timescale makes further predictions which BGK test with their data. As the burst ages, an increase in the strength of Hδ should signal the presence of A stars. Because the galaxies move apart as the burst ages, they should be closer to apogalacticon for larger EW(Hδ). At apogalacticon, the velocity difference of the pair is 0. Thus if we compare the distribution of line-of-sight velocity differences for galaxies with EW(Hδ)< -2 Å and EW(Hδ) ≥ 0, we should see a shift toward smaller velocity difference for EW(Hδ) absorption. We should see no relative shift in the ΔD distributions. Fig. 5 shows that the data confirm expectation.

If starbursts are tidally triggered, the burst should occur essentially simultaneously in both members of a pair. Fig. 6 shows the difference in EW(Hα) for galaxies in pairs (solid line) compared with a random pairing of EW(Hα)'s for galaxies in the sample. The observed distribution of differences in equivalent width is clearly shifted toward zero.

The large, objectively selected pair sample provides remarkably consistent support for many aspects of the MH96 picture of tidally triggered star formation. The MH96 n-body/SPH models suggest that tidal encounters can trigger central bursts of star formation in galaxies with shallow central po-

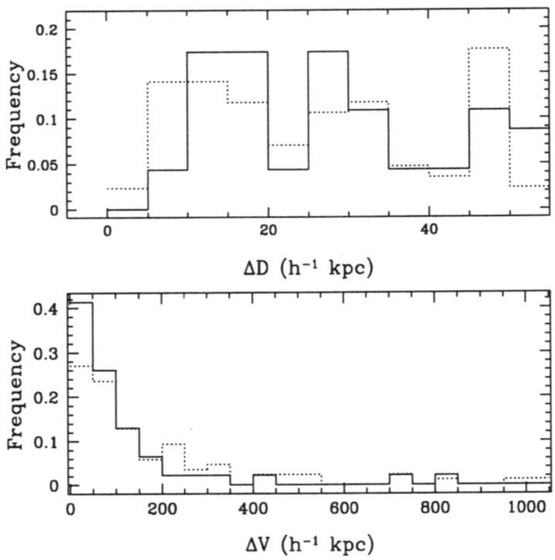

Fig. 5. Distribution of projected separations, Δ D, (upper panel) and line-of-sight velocity differences, Δ V, (lower panel) for galaxies with EW(Hδ)< -2 Å (solid line) and EW(Hδ)≥ 0 (dotted line).

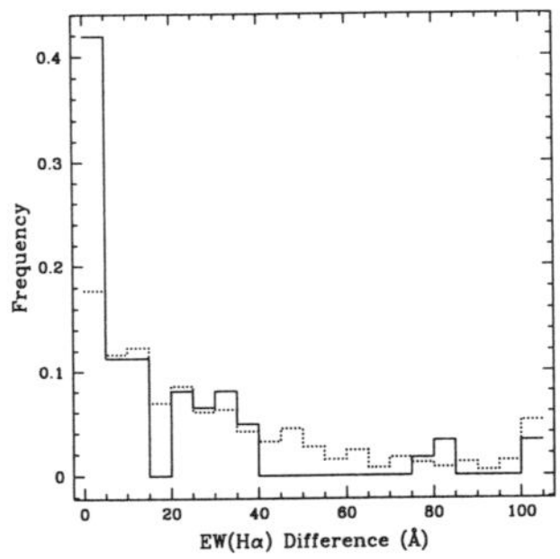

Fig. 6. Distribution of the difference in EW(Hα) for observed pairs (solid line) and randomly paired galaxies in the sample (dotted line).

tentials. Sophisticated as they are, these simulations do not treat star formation expressly; they show only that gravitational torques drive gas toward the center of a galaxy.

The bursts occur shortly after perigalacticon when the two galaxies are still separate systems. The central burst occurs whether the encounter is prograde or retrograde. For a subsample of their data with rotation curves, BGK find that the galaxies with the most intense central bursts do have shallow rotation curves. Furthermore central bursts occur in both prograde and retrograde encounters (see also Keel 1993 for a similar conclusion).

5. LARGE-SCALE STRUCTURE AND TIDALLY TRIGGERED STAR FORMATION

Physical explanations of the well-known morphology-density relation and the related dependence of spectroscopic type on local galaxy density are among the goals of large-scale simulations of the formation and evolution of the large-scale distribution of galaxies. Large-scale simulations which incorporate semi-analytic models for star formation are farther removed from the detailed physics than the simulations of MH96.

In these models, dissipationless n-body simulations track the formation and merging of dark matter haloes. A variety of phenomenological prescriptions and scaling laws normalized to observations account for gas cooling, supernova feedback and merging of galaxies within dark matter haloes. Standard models account for the quiescent evolution of the stellar population in each galaxy. On the grandest scales, this treatment yields an important result. By changing the prescriptions for star formation and feedback, different cosmological models provide consistent matches to the data displayed in Section 2 (Kauffmann 1999). This "degeneracy" warns that we must understand the objects we use to trace the universe if we want to understand the universe itself. In particular, we need a better understanding of star formation and all the associated processes if we want to understand galaxy formation and the large-scale structure we observe.

The cosmological n-body simulations with semi-analytic star formation carried out by Kauffmann et al (1999) show no correlation whatsoever between EW(Hα) and ΔD or ΔV on scales $\lesssim 50h^{-1}$ kpc even though, on larger scales, they track the behavior of population as a function of radius in clusters rather well. What is missing here, it seems, is the detailed, but perhaps crucial, sequence of tidally triggered starbursts during a merger.

The results of the simulations together with the observational results in Sections 2 and 3 suggest an explanation for the absence of a dependence of EW(Hα) in emission line galaxies on density smoothed over a large scale. If tidal triggering plays a major role, the EW(Hα) depends on the details of a relatively recent tidal encounter with a very near neighbor rather than on the global surroundings of a galaxy.

6. CONCLUSION

Large-scale surveys of the universe provide ways of probing the connections among physical processes which occur on vastly different scales. One example is the triggering of star formation by tidal encounters. The increasing quality and uniformity of the data along with large samples make these physical constraints possible. It is remarkable that our ability to observe and to simulate phenomena in the universe has advanced to a point where we can begin to make the kind of detailed physical models for interacting pairs of galaxies that Cecilia Payne-Gaposchkin made for interacting binary stars.

ACKNOWLEDGMENTS

We thank all of our collaborators Dan Fabricant, Scott Kenyon, Michael Kurtz and Brian McLean for their crucial contributions to the work described here. We also thank Paul Hodge for making us aware of Cecilia Payne-Gaposchkin's contributions to the understanding of star formation.

REFERENCES

Barton, E. J., Geller, M. J. & Kenyon, S. J. 2000 ApJ 530, 660

Barton, E. J., Geller, M. J. & Kenyon, S. J. 2001, in preparation

Carter, B. J., Fabricant, D. G., Geller, M. J., Kurtz, M. J. & McLean, B. 2001 ApJ, submitted

Dressler, A. 1980 ApJ 236, 351

Donzelli, C. J. & Pastoriza, M. G. 1997 ApJS 111, 181

Dubinski, J., Mihos, C. J. & Hernquist, L. 1996 ApJ 462, 576

Falco, E. E., Kurtz, M. J., Geller, M. J., Huchra, J. P., Peters, J., Berlind, P, Tokarz, S. & Elwell, B. 1999 PASP 111, 438

Geller, M. J. & Huchra, J. P. 1989 Science 246, 897

Geller et al. 2002, in preparation

Grogin, N. A. & Geller, M. J. 2000 AJ 113, 32

Hashimoto, Y., Oemler, A., Jr., Lin, H. & Tucker, D. L. 1998 ApJ 499, 589

Huchra, J. P. Geller, M. J., de Lapparent, V. & Corwin, H. G. Jr. 1990 ApJS 72, 433

Huchra, J. P., Geller, M. J. & Corwin, H. G., Jr. 1995 ApJS 99, 391

Kauffmann, G., Colberg, J. M., Diaferio, A. & White, S. D. M. 1999 MNRAS 303, 188

Keel, W. C. 1993 AJ 106, 1771

Kennicutt, R. C., Jr., Keel, W. C., van der Hulst, J. M., Hummel, E. & Roettiger, K. A. 1987 AJ 95, 5

Kochanek, C. S., Pahre, M. A. & Falco, E. E. 2001 astro-ph 0011458

Larson, R. B. & Tinsley, B. M. 1978 ApJ 219, 46

Leitherer, C., et al. 1999 ApJS 123, 3

Loveday, J., Tresse, L. & Maddox, S. 1999 MNRAS 310, 281

Mihos, J. C. & Hernquist, L. 1996 ApJ 464, 641

Postman, M. & Geller, M. J. 1984 ApJ 281, 95

Payne-Gaposchkin, C. 1952 Stars in the Making, Harvard University Press, Cambridge, p. 3

Sekiguchi, K. & Wolstencroft, R. D. 1992 MNRAS 255, 581

Toomre, A. & Toomre, J. 1972 ApJ 178, 623

Veilleux, S. & Osterbrock, D. E. 1987 ApJS 63, 295

York, D. and the Sloan Digital Sky Survey Collaboration 2000 AJ 120, 1579

Postman & Geller 1984

THE EVOLUTION OF THE MAGELLANIC CLOUDS: FROM CPG TO HST

Paul Hodge

University of Washington, Seattle, Washington

ABSTRACT: Using the distribution of Cepheid variables in the Magellanic Clouds, Cecilia Payne-Gaposchkin worked out a description of the spatial history of recent star formation in these galaxies. Her method has been rediscovered recently and, with extensions to other lines of evidence, especially cluster ages and deep color-magnitude diagrams, the method has become a powerful and widely-used technique. Her early results for the Magellanic Clouds are fully corroborated by modern results that use, for instance, the HST and the Zaritsky Great Circle Camera.

1. THE HARVARD VARIABLES

Early in the twentieth century Henrietta Leavitt and her Harvard colleagues identified over 2500 variable stars in the Magellanic Clouds. Many were classical Cepheids for which the Harvard astronomers defined the Period-Luminosity relation. But many remained unexplored and unexploited until the late 1950s, when the Gaposchkins began a major program of photometry and analysis.

Armed with the new data on magnitudes and periods of the Magellanic Cepheids, Prof. Payne-Gaposchkin realized their potential for exploring the recent history of star formation in the galaxy (Payne-Gaposchkin 1974). Arguing that the periods uniquely determine the ages of the Cepheids, she plotted their positions as a function of time, providing the first graphical history of a galaxy. Her plots of these data can be made into a time-lapse movie that illustrates the key conclusions that she made regarding the recent Magellanic Cloud evolution.

Fig. 1 reproduces an example of one of her plots of positions of the LMC Cepheids which she plotted as a sequence starting with the youngest (with longest periods) and ending with the oldest. The range in age is approximately 10^7 yrs to 10^8 yrs. For the shortest periods, data become incomplete, largely because these fainter Cepheids were not as easily discovered on the Harvard photographs.

She noted an obvious sequential star formation along the bar, where the most intense star-formation moves from the upper right to the lower left end of the bar.

The Starry Universe: The Cecilia Payne-Gaposchkin Centenary
A. G. Davis Philip & Rebecca A. Koopmann, eds.
p. 109-113

Paul Hodge
Department of Astronomy
University of Washington, Box 351580
Seattle, Washington 98195-1580
HODGE@ASTRO.WASHINGTON.EDU

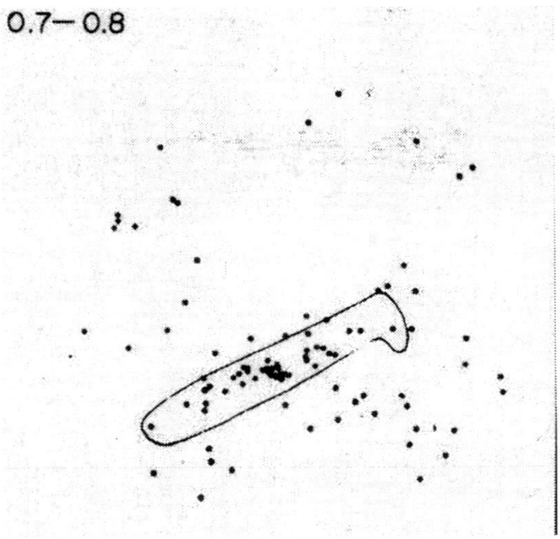

Fig. 1. One of Payne-Gaposchkin's (1974) plots of positions of LMC Cepheids, used by her to explore the recent history of star formation. This plots positions of Cepheids with log P between 0.7 and 0.9. Note the concentration of the middle of the bar.

The results of Payne-Gaposchkin's pioneering effort to trace Magellanic Cloud star formation both temporally and spatially can be summarized as follows:

a) The stars were formed in large-scale units about 1 kpc across. b) Star formation activity moved about over the face of the galaxies. c) In the case of the LMC, star-forming activity recently has moved NW along the bar.

These important new insights into the way stellar evolution proceeds across a galaxy have been corroborated in this century by modern surveys of variables in the Magellanic Clouds (see Freedman's essay in this volume).

2. THE LMC EVOLUTION FROM STAR CLUSTERS

Work at Harvard on the LMC star clusters also was laying the ground work for studying the star formation history of the galaxy. Annie Cannon had noticed that some of the LMC clusters had unusual spectra, A-type spectra instead of G or K. NGC 1866 (Fig. 2) was the brightest of these. Shapley and Nail (1949) discovered several classical Cepheids in this cluster and found that all had periods of about three days. This was later realized to be an indication that it must be a young cluster.

In Payne-Gaposchkin's 1974 analysis, she pointed to the importance of using star clusters to extend the range of time over which stellar formation events in the Magellanic Clouds could be traced. This was an approach that became the prime method for exploring the histories of resolved galaxies and continues to provide important insight.

Most subsequent study has concentrated on the massive, globular-like clusters of the LMC. A dozen of them are about 15 Gyr old and are virtually identical to our Galaxy's globular clusters (Olsen et al. 1998, Johnson et al. 1998). Fig. 3 shows NGC 1754, one of six such classical globular clusters in the LMC imaged with HST.

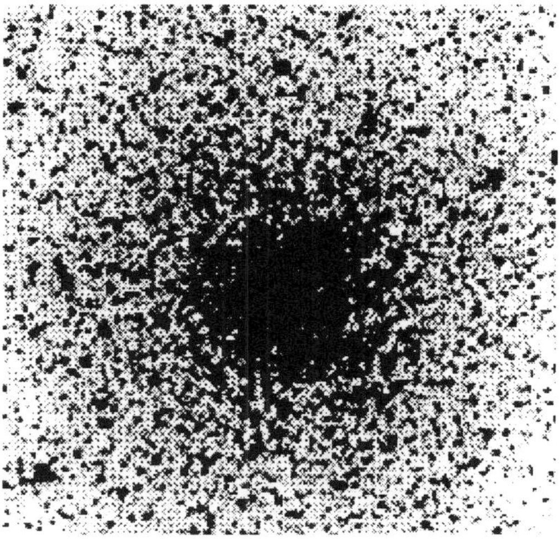

Fig. 2. The massive old (classical) globular cluster NGC 1754 in the LMC (NASA HST image).

But other populous clusters range in age down to only a few million years. NGC 2100 (Fig. 4) is a good example of a massive cluster that is only 20 million years old (Bohm-Vitense et al. 1985).

Fig. 3. The massive young star cluster NGC 1866 in the LMC (Boyden Observatory photo).

The distribution of ages of the massive clusters of the LMC show a very non-uniform formation rate. There seem to be peaks at 15 Gyr, at about 2 Gyr and 0.1 Gyr. These times appear to correspond closely to times when the LMC was interacting with the SMC and/or the Milky Way Galaxy.

3. THE LMC EVOLUTION FROM FIELD STARS

With HST it is now possible to derive star formation histories from Magellanic Cloud field stars, as well as clusters. For example, Olsen (1999) analyzed six LMC fields using HST images and found that field stars were formed sporadically over time. Differences seemed to be small and place-specific. As found for clusters, the peaks in star formation seemed to be roughly coincident with encounters between the Clouds and/or with the Milky Way, again suggesting tidally-induced enhanced star formation.

Similar conclusions have been reached for field stars in other parts of the LMC by Holtzman, Gallagher and many others using HST data (see Olsen 1999 for references) and by Harris and Zaritsky (Harris 2000) for both Magellanic Clouds using a huge survey made with the Great Circle Camera at Las Campanas Observatory

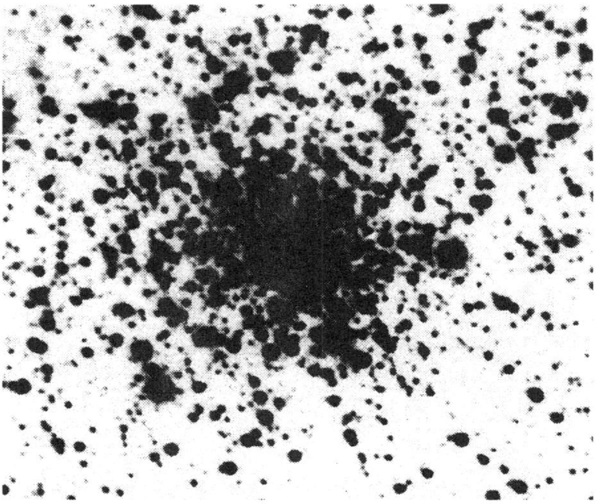

Fig. 4. The massive very young star cluster NGC 2100 in the LMC (Cerro Tololo Observatory photo).

4. CONCLUSIONS

Cecilia Payne-Gaposchkin pioneered a method of astronomical archeology, leading to an understanding of the history of the Magellanic Clouds.

This kind of research continues to provide us with important insight into the processes involved in the evolution of galaxies

REFERENCES

Bohm-Vitense, E., Hodge, P. & Proffitt, C. 1985 ApJ 292, 130

Johnson, J., Bolte, M., Stetson, P. B., Hesser, J. E. & Somerville, R. S. 1999 ApJ 527, 199

Harris, J. 2000 Ph.D. Thesis, University of California, Santa Cruz

Olsen, K. et al. 1998 MNRAS 300, 665

Olsen, K. 1999 Ph.D. thesis, U. of Washington

Payne-Gaposchkin, C. H. 1974 Smithsonian Contr. to Ap. 16, 1

Shapley, H. & Nail, V. 1949 Harvard Bull. No. 919

CEPHEIDS: FROM CECILIA TO THE PRESENT

Wendy L. Freedman

Carnegie Observatories, Pasadena, California

ABSTRACT: I briefly review some of Cecilia Payne-Gaposchkin's work on Cepheids in the Magellanic Clouds, as well as more recent work on Magellanic Cloud Cepheids, the properties of Cepheids in the context of their use as distance indicators, and progress in general on Cepheids and the extragalactic distance scale. In closing, I present the final results from the HST Key Project to measure the Hubble constant, based on the determination of distances calibrated using Cepheids.

1. PREAMBLE

Cecilia Payne-Gaposchkin died the year that I entered college as an undergraduate, so I unfortunately never had the opportunity to know her directly. I encountered her first simply through her and Sergei Gaposhkin's papers on Cepheids; only through her autobiography did I come to learn more about her. I recall at the time being particularly struck by Cecilia's description of attending a lecture by Sir Arthur Eddington on relativity. She was so excited by the talk that she returned to her room and recorded it, word for word. It was the first time I ever recalled someone describing the way I often felt after lectures while I was an undergraduate. No one I knew appeared to share the excitement I felt and that she was describing. I found in Cecilia's words, a certain rapport, and they have remained with me. So I am delighted to be taking part in this centenary celebration.

2. CEPHEIDS IN THE MAGELLANIC CLOUDS

Based on the Harvard plate collection, which was taken over a period of approximately 50 years, Cecilia and Sergei Gaposchkin undertook a systematic reanalysis of the variables in the Magellanic Clouds. Leavitt (1906) had discovered about 1,800 variables in the Magellanic Clouds; Shapley and collaborators continued the search thereafter. The period-luminosity relations for the LMC and SMC are reproduced below in Fig. 1 from Payne-Gaposchkin & Gaposchkin (1966). As noted in this reference, the slopes of the two PL relations are remarkably similar. In the case of the SMC, over 750,000 measurements were made on 500 plates taken with the 24-inch Bruce refractor from 1898

The Starry Universe: The Cecilia Payne-Gaposchkin Centenary
A. G. Davis Philip & Rebecca A. Koopmann, eds.
p. 115-122

Wendy L. Freedman
Carnegie Observatories
813 Santa Barbara St.
Pasadena, California 91101
WENDY@OCIW.EDU

to 1962 (Payne-Gaposchkin & Gaposchkin 1966). Sergei made, or supervised, the measurements [1] while Cecilia measured the periods. Examples of some of the light curves are reproduced in Fig. 2.

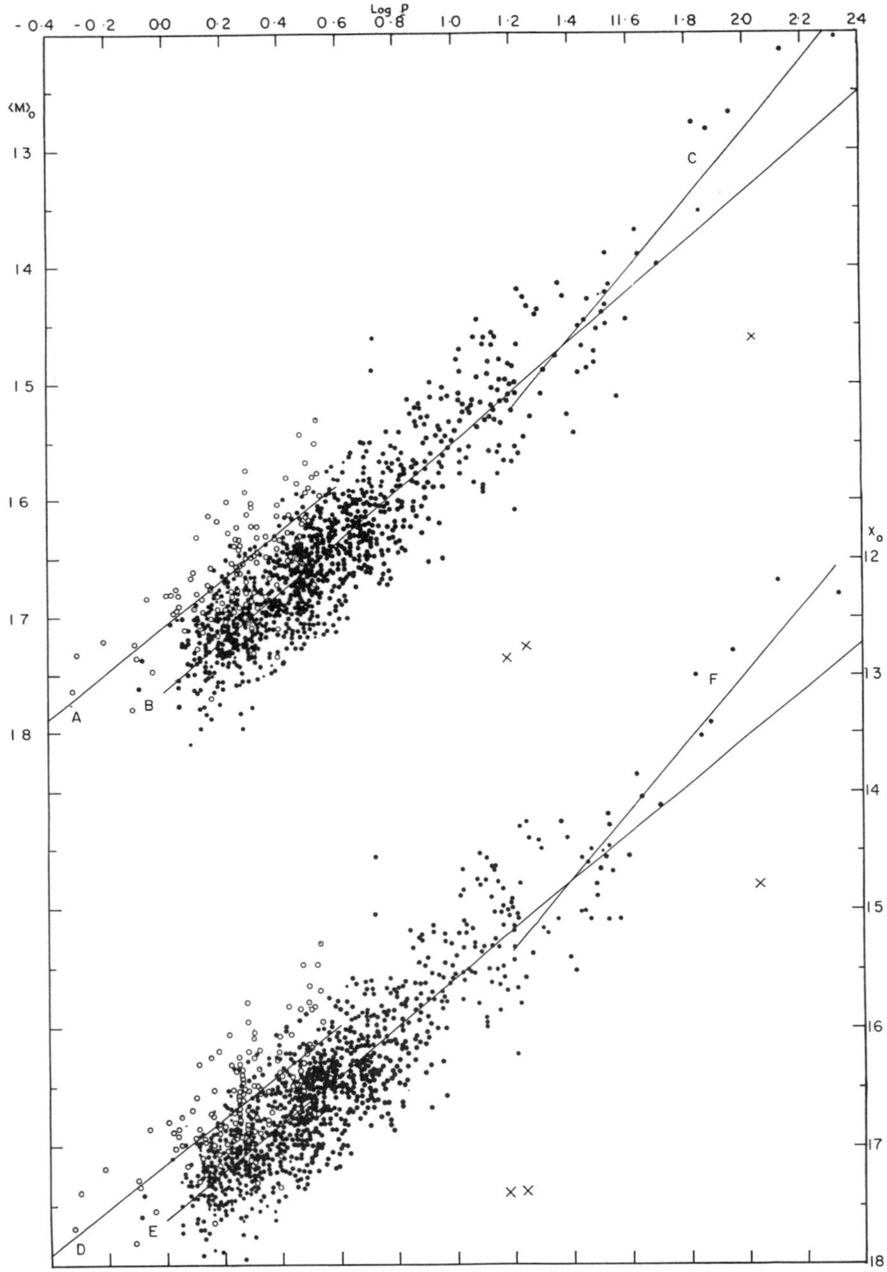

Fig. 1. Period-luminosity relations for the LMC and SMC (Payne-Gaposchkin 1974)

[1] I learned at this meeting that as a student, Ken Janes made some of these measurements.

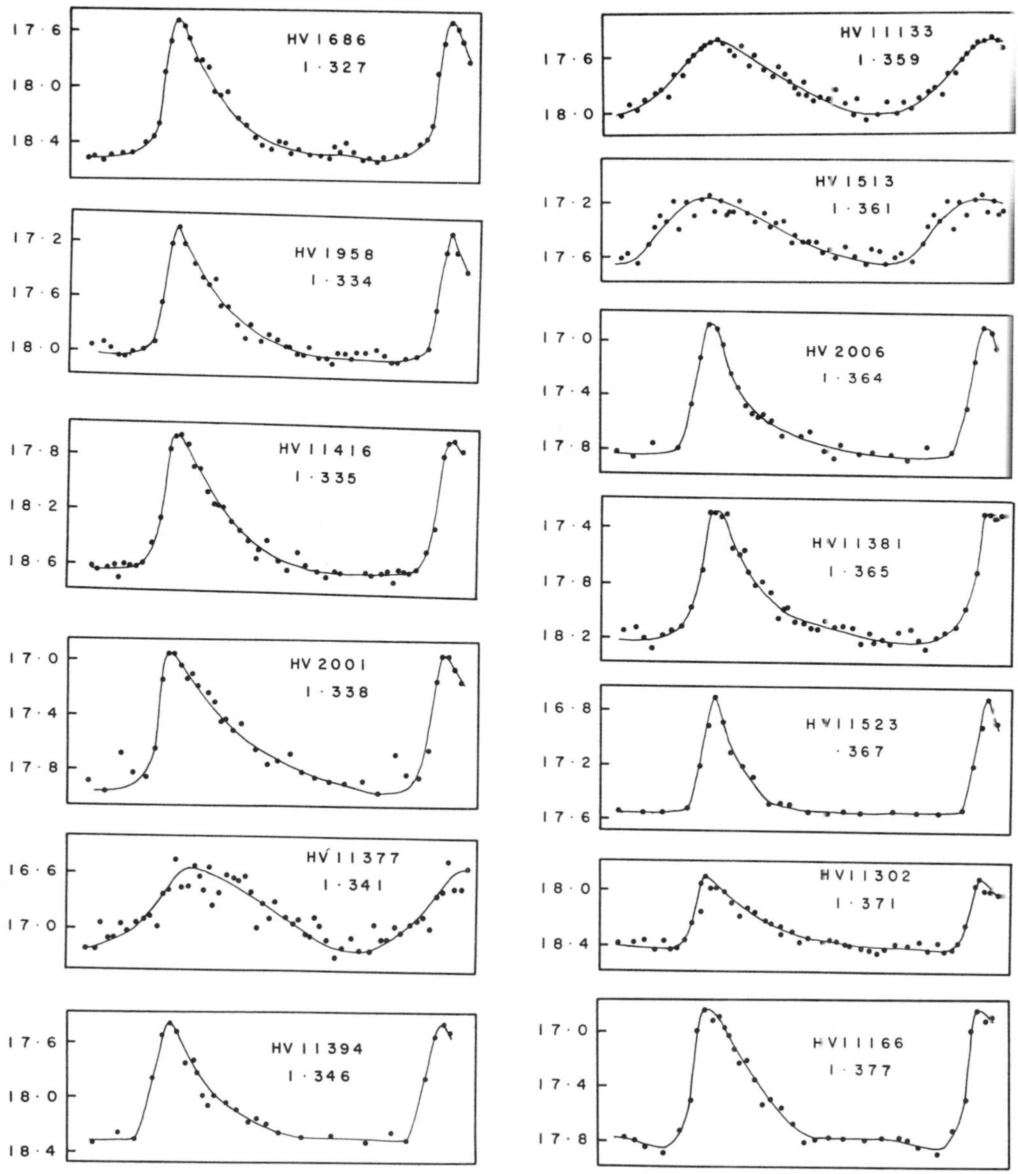

Fig. 2. B-band light curves for Cepheids in the SMC (Payne-Gaposchkin & Gaposchkin 1966)

Note the excellent quality and low scatter in these data.

Using the correlation between the period of a Cepheid and its age, Payne-Gaposchkin (1974) later outlined the broad history of recent star formation in the LMC. She also noted the distribution of the shortest-period Cepheids in the southern end of the bar, and the progression to longer periods (younger stars) toward the northern end of the bar. The period distributions are reproduced in Fig. 3. The discovery of new Cepheids in the Magellanic Clouds has been recently energized as a byproduct of several independent searches for microlensing from massive compact halo objects (MACHOS).

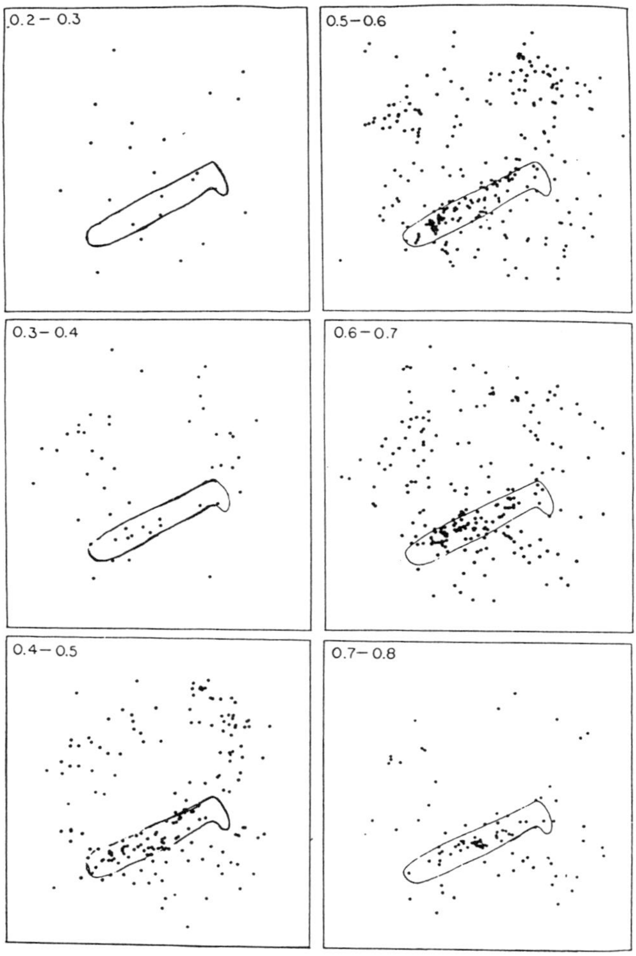

Fig. 3. Distribution of shorter-period Cepheids in the LMC. The logarithm of the period is shown in the upper left of each panel. The LMC bar is outlined (Payne-Gaposchkin 1974).

Alcock et al. (1999) have discovered about 1,800 Cepheids in the LMC (551 of which were part of the Payne-Gaposchkin sample); Udalski et al. (2000) have discovered about 1,300. Alcock et al. confirm the overall conclusions of Payne-Gaposchkin regarding the star formation history of the LMC; the burst of star formation about 100 million years ago (with a duration of about 20 million years), and propagation of star formation from the southeast toward the northwest along the bar They conclude that it took about 40 million years for star formation to proceed from one end of the bar to the other.

3. CEPHEIDS AS DISTANCE INDICATORS

An historical review on the use of Cepheids as distance indicators can be found in Fernie (1969). Beginning in the 1950's, Cepheids in galactic clusters were used as a means of independently setting the Cepheid period-luminosity relation zero point, using Pop I main-sequence fitting. Prior to the 1980's, the most serious systematic error embedded in the Cepheid distance scale concerned the assumption that the Cepheids in external galaxies suffered no absorption internal to the parent galaxies in which the Cepheids were being studied. This assumption was eventually shown to be false, largely as a result of the availability of new detectors over the past two decades (CCDs, IR photometers and arrays). More recent reviews of the Cepheid distance scale include those of Jacoby et al. (1992) and Madore & Freedman (1991). In the latter, the basic physics underlying the Cepheid period-luminosity (PL) and period-luminosity-color (PLC) relations, the effects of reddening, metallicity and photometric errors are reviewed.

Cepheids have remained the most accurate primary method for measuring distances to external galaxies. The periods of classical Cepheids range from approximately two to over 100 days. Cepheids are characterized by their distinctive light curves having a rapid rise to maximum light, followed by a slower linear decline to minimum light. Below a period of about nine days Cepheids are observed to pulsate in both the fundamental, in addition to the first harmonic (overtone) mode. Cepheids have absolute visual magnitudes in the range $-2 < M_V < -7$ mag (corresponding to $\sim 10^3$ - 10^5 solar luminosities), and supergiant spectral types ranging from early F to late K. The amplitudes of all Cepheid variables decrease toward redder wavelengths. The large relative amplitudes at blue and ultraviolet wavelengths are due to the higher sensitivity of monochromatic surface brightness to temperature variations at short wavelengths; whereas the low amplitudes in the red and near-infrared are primarily controlled by the small changes in surface area driven by the radial pulsation. Cepheid searches are therefore most efficiently carried out at short wavelengths. Observations at longer wavelengths are needed to allow for corrections due to interstellar reddening and metallicity.

As distance indicators, the strengths of Cepheids are many: they are amongst the brightest stellar indicators, and they are relatively young stars, being found in abundance in spiral galaxies. Thus, many independent objects can be observed in a single galaxy. Their large amplitudes and characteristic (saw-tooth) light curve shapes facilitate their discovery and identification; and they have long lifetimes and hence, can be reobserved at other times and/or at other wavelengths. The Cepheid relation has a small scatter (e.g., in the I-band, the dispersion amounts to only $\sim \pm 0.1$ mag). Moreover, Cepheids have been studied and theoretically modeled extensively; the reason for their variability is well-understood to be a consequence of pulsation of the atmosphere, resulting from a thermodynamic, valve-like driving mechanism as helium (primarily) is cycled from a singly to doubly ionized state, and the opacity increases with compression.

There are also difficulties associated with measuring Cepheid distances. First, since Cepheids are young stars, they are found in regions where there is dust scattering, absorption and reddening. Corrections must be made for extinction, requiring assumptions about the universal behavior of Cepheids at different wavelengths, and about the universality of the galactic extinction law. Extinction is systematic, and its effects must either be removed by multicolor data, or minimized by observing at long wavelengths, or both. Second, the dependence of the PL relation on chemical composition (metallicity) has been very difficult to quantify. Third, an accurate geometric calibration of the PL relation, at any given metallicity, has not yet been established. Fourth, as the distance of the galaxy increases (and the resolution decreases), finding and measuring individual Cepheids becomes increasingly difficult due to crowding effects. Finally, the reach of Cepheids is currently (with HST) confined to spiral galaxies with distances less than about 30 Mpc.

At the time of this writing, Cepheids have now been surveyed for and discovered in over 40 galaxies. The numbers have rapidly increased over the past decade as additional galaxies have been surveyed with the HST. The numbers of Cepheids in the Magellanic Clouds has also been increased due to the intensive searches carried out for microlensing from massive compact halo objects by the MACHO (wwwmacho.anu.edu.au/), OGLE and EROS (www.lal.in2p3.fr/recherche/eros/erosa.html) collaborations.

One of the primary motivations for building the Hubble Space Telescope (HST) was to enable the discovery of Cepheids in galaxies as distant as the Virgo cluster, with the ultimate goal of measuring the Hubble constant (H_0) to higher precision than could be done from the ground. Cepheids alone cannot be observed at sufficient distances to determine H_0 directly, and an accurate determination of H_0 requires an extension to other methods. As part of the H_0 Key Project (Freedman et al. 2001, Mould et al. 2000) Cepheid distances were obtained for 18 galaxies with distances between three and 25 Mpc to provide a calibration for several, independent secondary methods extending out to distances of 50 - 400 Mpc. HST has also been used to measure Cepheid distances to six additional galaxies, targeted specifically to be useful for the calibration of Type Ia supernovae (e.g., Sandage et al. 1996), and several other Cepheid distances have been measured recently by a number of other groups. For each galaxy observed as part of the Key Project, the Cepheid positions, magnitudes, and periods are available at http://www.ipac.caltech.edu/H0kp/H0KeyProj.html. In addition, photometry for non-variable stars that can be used for photometry comparisons, as well as medianed (non-photometric) images for these galaxies are also available. These images are also archived in NED, and can be accessed on a galaxy-by-galaxy basis from http://nedwww.ipac.caltech.edu.

4. THE HUBBLE CONSTANT AND EXPANSION AGE

Calibrating five secondary methods with Cepheid distances, Freedman et al. (2001) find $H_0 = 72 \pm 3$ (random) $\pm$ 7 (systematic) units. Type Ia supernovae are the secondary method which currently extends out to the greatest distances, ~400 Mpc. All of the methods (Types Ia and II supernovae, the Tully-Fisher relation, surface brightness fluctuations, and the fundamental plane) are in extremely good agreement: four of the methods yield a value of H_0 between 70 - 72 units, and the fundamental plane gives $H_0 = 82$ units. As described in detail in Freedman et al., the largest remaining sources of error result from (a) uncertainties in the distance to the Large Magellanic Cloud, (b) photometric calibration of the HST Wide Field and Planetary Camera 2, (c) metallicity calibration of the Cepheid period-luminosity relation, and (d) cosmic scatter in the density (and therefore, velocity) field that could lead to observed variations in H_0 on very large scales.

For many years, there has been concern that larger values of the Hubble constant lead to a discrepancy with the measured ages of stars in galactic globular clusters. Recent parallax measurements from HIPPARCOS have resulted in a calibration yielding lower ages for globular clusters of 12 - 13 Gyr (Chaboyer 1998, Carretta et al. 2000). For a value of $H_0 = 72$ units, a flat, matter-dominated universe with $\Omega_m = 1$ (and no cosmological constant) results in a very young expansion age of only 9 ± 1 Gyr, still significantly younger than the globular cluster and other age estimates. A non-zero value of the cosmological constant helps to avoid a discrepancy between the expansion age and other age estimates. For $H_0 = 72$ units, $\Omega_m = 0.3$, $\Omega_\Lambda = 0.7$, the expansion age is 13 ± 1 Gyr, consistent to within the uncertainties, with recent globular cluster ages, and also with measurements of distant Type Ia supernovae (e.g., Riess et al. 1998, Perlmutter et al. 1999) and the cosmic microwave background (de Bernardis et al. 2000, Lange et al. 2000).

ACKNOWLEDGMENTS

I thank the organizers of this meeting for the opportunity to speak at one of the most enjoyable conferences that I have attended. I learned much about Cecilia's scientific contributions at this meeting. There was much to celebrate on this occasion of her centenary. An unexpected pleasure for me, however, was getting to know more about Cecilia informally during coffee breaks and meals, from people who had known her during her life. I also would like to thank my many collaborators on the extragalactic distance scale over the years, most recently those involved in the HST Key Project, and especially Barry Madore.

REFERENCES

Alcock, C., et al. 1999 AJ 117, 920

Carretta, E., Gratton, R., Clementini, G. & Fusi Pecci, F. 2000 ApJ 533, 215

Chaboyer, B., Demarque, P., Kernan, P. J. & Krauss, L. M. 1998 ApJ 494, 96

de Bernardis, P., et al. 2000 Nat. 404, 955

Fernie, J. D. 1969 PASP 81, 707

Freedman, W. L., et al. 2001 ApJ, Apr. 10, in press, astro-ph/0012376

Jacoby, G. H., et al. 1992 PASP 104, 599

Lange, A. E., et al. 2000 astro-ph/0005004

Leavitt, H. S. 1906 Ann. Harvard Coll. Obs, 60, 87

Madore, B. F. & Freedman, W. L. 1991 PASP 103, 933

Mould, J. R., et al. 2000, ApJ 529, 786

Payne-Gaposchkin, C. & Gaposchkin, S. 1966 Smithsonian Contr. Ap., Vol. 9

Payne-Gaposchkin, C. 1974 Smithsonian Contr Ap. 16

Perlmutter, S., et al. 1999 ApJ 517, 565

Riess, A. G., et al. 1998 AJ 116, 1009

Sandage, A. R., Saha, A., Tammann, G. A., Labhardt, L., Panagia, N. & Macchetto, F. D. 1996 ApJ 460, 15

Udalski, A., Soszynski, I., Szymanski, M., Kubiak, M., Pietrzynski, G., Wozniak, P. & Zebrun, K. 1999 Acta Astronomica 49, 223

SYMBIOTIC BINARY STARS

Scott J. Kenyon

Smithsonian Astrophysical Observatory, Cambridge, Massachusetts

ABSTRACT: Symbiotic stars are interacting binary stars with a red giant primary star, a hot secondary (main-sequence, white dwarf, or neutron star), and an ionized nebula. This paper briefly reviews the early discovery of symbiotic stars, modern observations of one of the first catalogued symbiotic stars, AG Pegasi, and an interesting symbiotic cousin, BG Gem, which may contain a black hole.

1. INTRODUCTION

The theory of stellar evolution is one of astronomy's greatest achievements. In the past century, advances in physics and numerical computation have led to a detailed understanding of the time evolution of single and multiple stars. By discovering that stars are mostly hydrogen, Cecilia Payne established the cornerstone for this theory. Her work on the astrophysics of variable stars - including Cepheid variables, classical novae, and supergiant stars - tested the theory at its extremes, where our understanding of stellar physics is weakest. Without these contributions, the theory of stellar evolution would have been much less rich and would have progressed much more slowly.

In "The Galactic Novae", Payne-Gaposchkin emphasized that acceptable theories depend on a deep understanding of the facts and a thorough knowledge of astrophysics and cosmogony. The symbiotic stars demonstrate her wisdom. During the compilation of the HD Catalog, Fleming (1907) identified many late-type stars with strong emission lines. Merrill (1944) divided these stars into five groups based on the visibility of *specific* emission lines, such as H I or Fe II. Merrill's third class - stars with combination spectra - eventually became known as the symbiotic stars. When Payne-Gaposchkin wrote "The Galactic Novae", she noted the many similarities between *symbiotic novae,* as she called them, and classical novae. Despite the lack of consensus on their binary nature, she noted that "a third . . . may be considered to be binary objects. That all of them are binary may be used as a working hypothesis to be tested by exploring its implications". Together with work by Merrill and Swings & Struve, chapter 7 of "The Galactic Novae" laid out the implications of the binary hypothesis, where

The Starry Universe: The Cecilia Payne-Gaposchkin Centenary
A. G. Davis Philip & Rebecca A. Koopmann, eds.
p. 123-134

Scott J. Kenyon
Smithsonian Astrophysical Observatory
60 Garden Street
Cambridge, MA 02138
SKENYON@CFA.HARVARD.EDU

a red giant and a hot companion produce the combination spectrum. Radial velocity measurements of the absorption spectrum should yield the orbital motion of the red giant primary star. Without ultraviolet spectra to detect the continuum from a hot star, radial velocity variations of strong emission lines might reveal the motion of the hot companion to the giant. It took 25 years to collect the right facts to test the binary hypothesis. Spectra from the International Ultraviolet Explorer - launched shortly before Payne-Gaposchkin's death - detected a strong ultraviolet continuum from every bright symbiotic star (Kenyon 1986). Spectrographs equipped with Reticon and CCD detectors later revealed orbital motion. These and other observations confirmed the working hypothesis that symbiotic stars are binary stars, similar in many respects to cataclysmic variables and Algol systems. Once these facts were established, additional theoretical work showed that symbiotic stars lie at an important crossroads in the evolution of interacting binaries, where some systems evolve into cataclysmic variables and others become type Ia supernovae (see Kenyon 1986). Payne-Gaposchkin's outline to test the binary hypothesis has now moved from testing the implications of a binary symbiotic star to testing the implications of the evolution of an ensemble of binary symbiotic stars.

In the rest of this paper, I review recent observations of a well-known symbiotic, AG Pegasi, and a cousin to the symbiotic stars, BG Geminorum. The symbiotic nova AG Peg shows how modern data yield a definitive test of the binary hypothesis for all symbiotics. The shorter period binary BG Gem shows how a new discovery can add to our collection of facts and perhaps change our understanding of stellar evolution theory.

2. AG PEGASI: THE SLOWEST CLASSICAL NOVA EVER

The symbiotic binary AG Peg began a slow nova eruption in the 1850's, rising from $V \sim 9$ to $V \sim 6$ in $\sim$ one decade (Lundmark 1921). Spectra c. 1900 revealed a peculiar Be-type spectrum with H I emission lines and absorption from He I and TiO (Fleming 1907, Merrill 1916, Payne-Gaposchkin 1957). Merrill (1929a, b) later identified an $\sim$ 800 day periodicity in the emission line intensities, and in the radial velocities of the absorption and the emission lines. Over the next two decades, the absorption line velocities continued to follow this period; the relative phasing of the emission lines changed by $\sim$ 200 days. Most of the observations supported the binary hypothesis, but the remarkable changes in the emission lines made it difficult to locate the 'hot component' relative to the red giant.

The optical spectrum of AG Peg has changed dramatically since 1915 (Fig. 1). The slow evolution of the hot component from a Be-type star with narrow emission lines into a Wolf-Rayet star with broad emission lines took $\sim$ 30 years. The blue continuum declined by 0.5 - 1.0 mag (Payne-Gaposchkin 1952). As the broad Wolf-Rayet features appeared, Merrill (1959) noted that the intensities and radial velocities of [Ne III] and [O III] λ 4363 followed the orbital period, while those of [O III] $\lambda\lambda$ 4959, 5007 did not. After analyzing the differences in intensities, line profiles, and radial velocities, Merrill (1959) proposed "a definitely directed stream of gas", to produce λ 4363 emission near the central binary and to feed a larger, expanding volume of gas emitting mostly λ 5007. Merrill adopted Payne-Gaposchkin's term, "the fire hose effect", to describe the ejection of material from the AG Peg binary. In my search of the literature using the ADS Abstract Service, the jet in AG Peg is the first ever known in a binary star system, discovered at roughly the same time as the jet in M 87 (Baade 1956).

Despite the remarkable changes in the hot component, the red giant in AG Peg remained constant. Better definition of the red giant absorption features as the blue continuum from the hot component declined yielded a better estimate of its spectral type, M2 or M3, instead of the preliminary M1 type. From all published accounts, the red magnitude of the giant did not change during the 1 mag decline of the hot component.

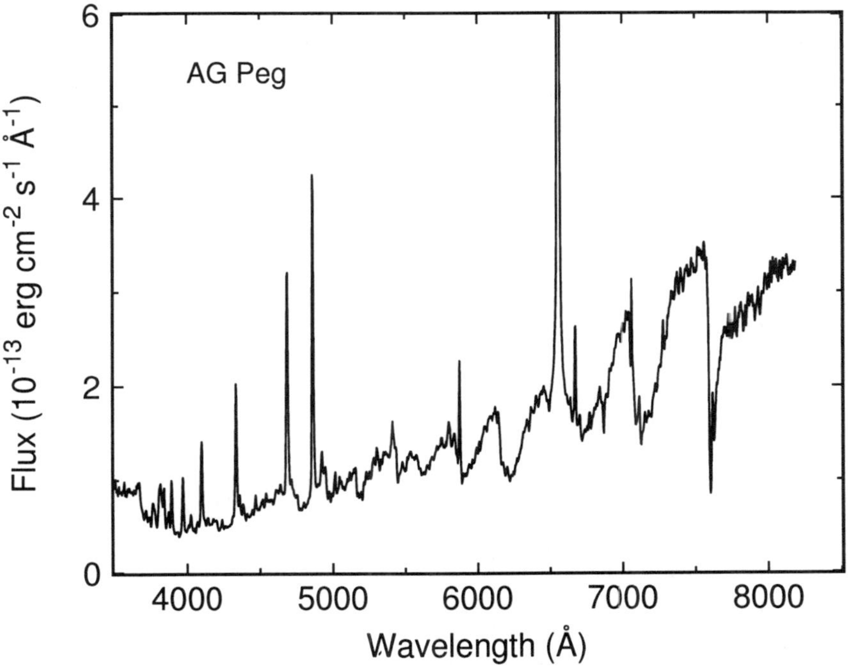

Fig. 1. Spectrum of AG Peg during its decline from the c. 1850 outburst from Kenyon et al. (1993). The red end of the spectrum is dominated by TiO absorption bands and a strong Hα emission line. The blue end of the spectrum shows a weak continuum and many strong emission lines.

These observations led to a simple picture for the geometry of AG Peg, where an M-type giant and the unstable hot component lie embedded in a common nebulosity fed by a "fire hose" or "jet". Merrill and Belyakina (1968, 1970) suggested that narrow emission from Fe II and other low ionization lines forms in the heated hemisphere of the red giant; broad, high ionization emission lines form near the hot component. The slow increase in ionization with time might be an increase in the temperature of the hot component, but the origin of this increase was unknown. The origin of the jet and emission from the red giant photosphere was also unclear. Phase lags in the emission lines relative to the orbit of the M giant made it difficult to understand the relation of the broad lines to the jet.

Ultraviolet data - beginning with ANS, OAO-2, and TDS-1 and continuing with IUE and HST - demonstrate conclusively that AG Peg is an interacting binary star. The low resolution ultraviolet spectrum from the Hubble Space Telescope in Fig. 2 has a Rayleigh-Jeans tail from a hot continuum for $\lambda \lesssim 2000$ Å, a flat continuum for $\lambda \approx 2000$ - 3646 Å, and a red continuum from the cool giant for $\lambda \gtrsim 4000$ Å. Strong emission lines from a wide range of ionization states are superimposed on these continua. The strong ultraviolet continuum and emission lines imply effective temperatures of 10^5 K or more for the hot white dwarf in AG Peg (Kenyon et al. 1993, 2001). The hot components in other symbiotics are usually very hot white dwarfs, sometimes an accreting main-sequence star, and in one case a neutron star (Kenyon & Webbink 1984, Kenyon 1986, Mürset et al. 1991).

Together with optical spectra, these ultraviolet data yield a good picture for the evolution of the hot component in AG Peg. If the B-type spectrum can be associated with a stellar photosphere, the hot component has become steadily hotter over the past 100 years, from 100,000 K $\sim$ 10,000 K c. 1900 to $\gtrsim$ 100,00 in 1995 (Kenyon et al. 1995, 2001). The optical continuum from the hot component fell in step with the increase in effective temperature. Assigning crude bolometric

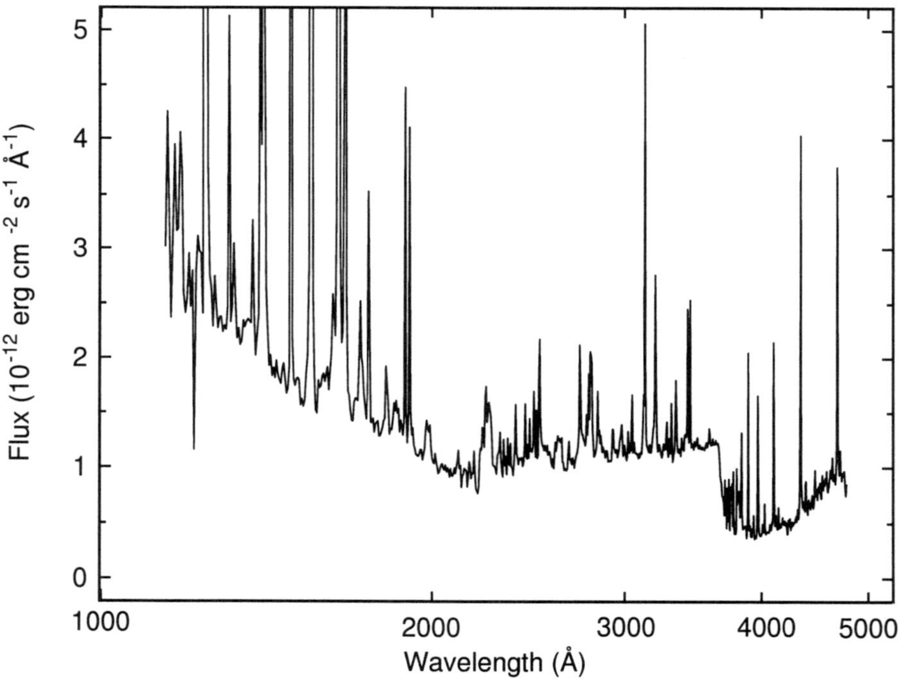

Fig. 2. HST spectrum of AG Peg from Kenyon et al. (2001). The data have not been corrected for interstellar reddening. The continuum consists of the Rayleigh-Jeans tail of a hot blackbody for $\lambda < 2000$ Å. Strong nebular emission for 2000 Å$< \lambda < 3646$ Å, and the Wien side of the red giant spectrum for $\lambda > 4200$ Å. Intense permitted and forbidden emission lines are visible at all wavelengths.

corrections to the optical spectra at several epochs, Gallagher et al. (1979) and Kenyon et al. (1993, 2001) show that the hot component maintained a nearly constant luminosity during most of its remarkable 100 year evolution (Fig. 3).

The constant luminosity of the hot component in AG Peg resembles the constant luminosity phases observed in slow classical novae, such as FH Ser or HR Del (Gallagher & Code 1974, Gallagher & Starrfield 1978). Slow classical novae require $\sim$ 1 - 3 yr to return to quiescence instead of more than 100 yr. Classical novae are more luminous, $\gtrsim$ 10,000 $L_{\odot}$, than the hot component in AG Peg, $\sim$ 3,000 $L_{\odot}$. Nevertheless, detailed stellar evolution calculations show that the thermonuclear runaway model which successfully explains classical nova outbursts can also account for the eruption of AG Peg, providing the system is a wide binary containing a relatively low mass white dwarf (Kenyon & Truran 1983). To make the observational distinctions clear, Allen (1980) called AG Peg, RR Tel and other symbiotics with slow nova-like behavior *very slow novae*. In honor of "The Galactic Novae", Kenyon & Truran (1983) called them symbiotic novae, the term most commonly used today.

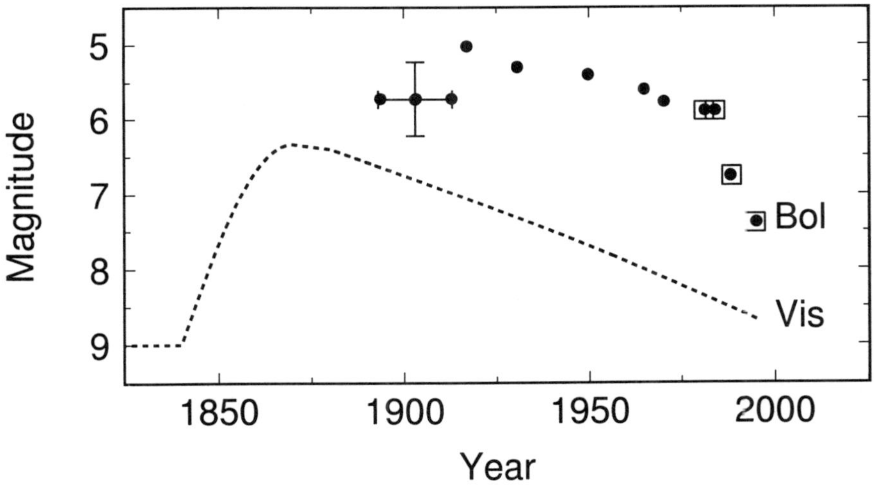

Fig. 3. Variation of optical (dashed line) and bolometric luminosity (symbols) of the hot component with time. The open circles are luminosities from Kenyon et al. (1993) using bolometric corrections estimated from optical spectra. The cross indicates a typical error bar for these estimates. The open boxes surrounding filled circles are luminosities derived from UV data in Kenyon et al. (1993, 2001).

Modern digital spectra also yield a better understanding of the M3 giant star in AG Peg. The strengths of the TiO and VO features on optical spectra vary in phase with the 816 day orbital period. The bands are weaker when we view the red giant hemisphere facing the hot component, and stronger when we view the opposite hemisphere. The intensities of the H I and He I emission lines vary 180° out of phase with the absorption lines. Both of these facts indicate that radiation from the hot component heats the atmosphere of the red giant, producing emission lines in the outer atmosphere and weakening the absorption features formed in deeper layers. To test this picture in detail, Proga et al. (1996, 1998) calculated radiative transfer models for red giants illuminated by radiation from a nearby hot white dwarf. These models show that emission from a static red giant atmosphere fails to reach the observed levels by 1 - 2 orders of magnitude. An illuminated red giant wind can explain observed emission line fluxes if the mass loss rate in the wind is $\sim 10^{-7}$ $M_\odot$ yr^{-1}. This rate agrees with estimates derived from radio observations of the ionized nebula.

Our understanding of ionized material outside of the red giant atmosphere is still poor. Merrill's "jet" of gas has evolved considerably. Although [O III] and [Ne III] features are now extremely weak or absent, VLA images reveal an unresolved central source embedded in several extended shells of gas (Kenny et al. 1991). The smallest of the extended radio sources is probably the expanding [O III] λ 5007 region Merrill (1959) identified on high resolution spectra. This region produces most of the 1 - 20 cm emission; it has a radius of $\sim$ 400 - 500 AU and is expanding at a rate of $\sim$ 60 km s^{-1} if the distance to AG Peg is 600 pc (Kenny et al. 1991). The expansion rate agrees with rates measured from the [O III] lines in the 1950's (Merrill 1959) and the 1970's (Hutchings, Cowley & Redman 1973). The size of this region is also consistent with an expansion starting in 1940 - 50, as proposed by Merrill (1959).

The radio nebula may also be a source of high ionization optical forbidden line emission. Lines of [Ne V] and [Fe VII] appeared in the mid-1980's and have remained at roughly constant intensities for a decade. The relative line strengths suggest formation in a hot, dense gas with an electron temperature of $\sim$ 100,000 K and an electron density of $\sim 10^7$ cm^{-3}. The large emission measures for Ne^{+4} and Fe^{+6} gas preclude formation close to the hot component or immediately surrounding the red giant. The modest density favors material in the expanding radio shell, but the hot component is too cool to ionize Ne or Fe as observed. Shocks in the red giant wind are a potential excitation mechanism for this gas, but the kinetic energy in the winds from the hot component and the red giant are smaller than the emitted energy by a factor of $\sim$ 10 or more.

To test possible models for the shocked gas, we can follow Payne-Gaposchkin's maxim. At a distance of more than $0\farcs5$ from the central binary, the inner radio nebula can be resolved with STIS on board HST. These observations would test the notion that the forbidden emission lines are formed well outside the red giant atmosphere. Resolving the nebulosity within the central binary, $\lesssim 0\farcs005$ across, is more difficult but may be possible with ground-based interferometers or future space missions. Higher spectral resolution UV or optical observations would yield line profiles for the high ionization lines and provide better constraints on the shock velocity and geometry.

3. BG GEMINORUM: A POSSIBLE BLACK HOLE

For over 50 years, BG Gem languished in The General Catalog of Variable Stars as a possible pulsating star with an uncertain period of 60 days. In 1992, P. Benson and her students began photometric observations with the Wellesley College 24″ Sawyer telescope and CCD camera. Their first year of observations revealed a repeatable sinusoidal variation with a period of roughly 90 days. With a summer 1998 grant from the Keck Northeast Astronomy Consortium, A. Bonanos compiled and analyzed six years of CCD observations taken through broadband Kron-Cousins VRI filters. Her results demonstrated that BG Gem is an eclipsing binary with a 91.7 day orbital period. The following year, A. Dullighan - also using a grant from the Keck Northeast Astronomy Consortium - refined the orbital period to 91.645 days.

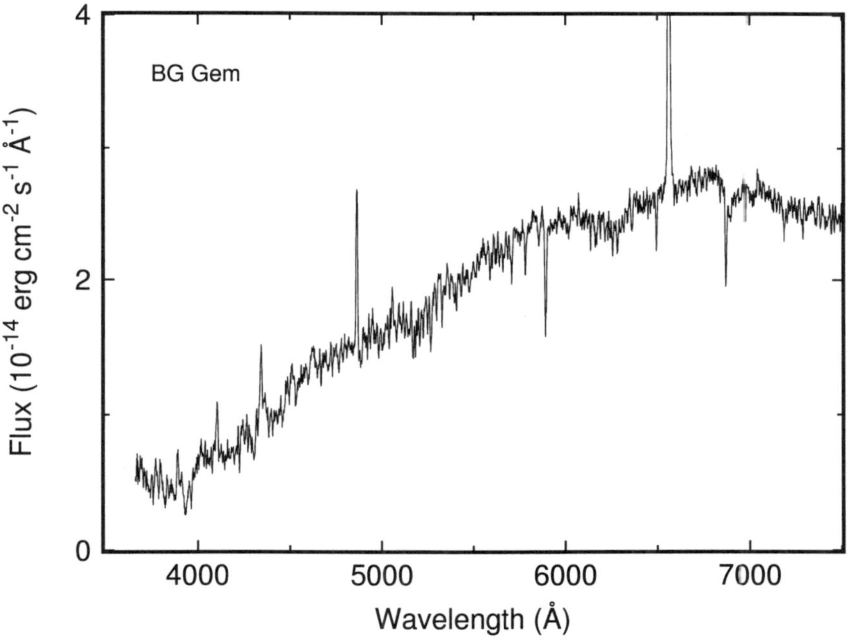

Fig. 4. Optical spectrum of BG Gem (Benson et al. 2000). Strong H I emission lines and a faint blue continuum shortwards of 4000 Å are produced by the hot primary. The red continuum and prominent Mg I and Na I absorption lines are from the K0 I secondary star.

To learn more about the eclipses in BG Gem, P. Benson, K. McLeod and I acquired spectra with the FAST spectrograph designed by D. Fabricant for the 60" Tillinghast telescope at Whipple Observatory. Fig. 4 shows a spectrum near maximum light. Aside from telluric absorption lines at 6900 Å, this spectrum shows strong Na I absorption lines at 5900 Å, a strong blend of Ba I and other lines at 6495 Å, and many other weaker features. The relative strengths of these lines indicate a K0 I spectral type. Strong H I emission lines and a weak blue continuum are the only optical evidence for a hotter companion star.

Fig. 5 shows optical light curves for BG Gem. Ellipsoidal variations are prominent. Light curve calculations using the Wilson-Devinney code (Wilson & Devinney 1971, Wilson 1994) imply that a lobe-filling K-type secondary star emits most of the optical light. At 4000 Å, the deep primary minimum at ϕ = 0.0 requires the eclipse of an extended primary star. This primary star occults the K-type secondary star during deep eclipses at R_C and I_C at ϕ = 0.5. The solid lines in Fig. 5 plot model light curves from the Wilson-Devinney code assuming the primary is an accretion disk. A disk with scale height, H_d = 0.05 - 0.1 R_d, where R_d is the disk radius, allows reasonable fits to the data. An upper limit to the mass ratio is $q = M_2/M_1 = 0.1 \pm 0.05$ for sin i = 1. The eclipses constrain the orbital inclination to be $i \gtrsim 80°$. Detection of optical radiation from the disk requires $i \lesssim 88°$.

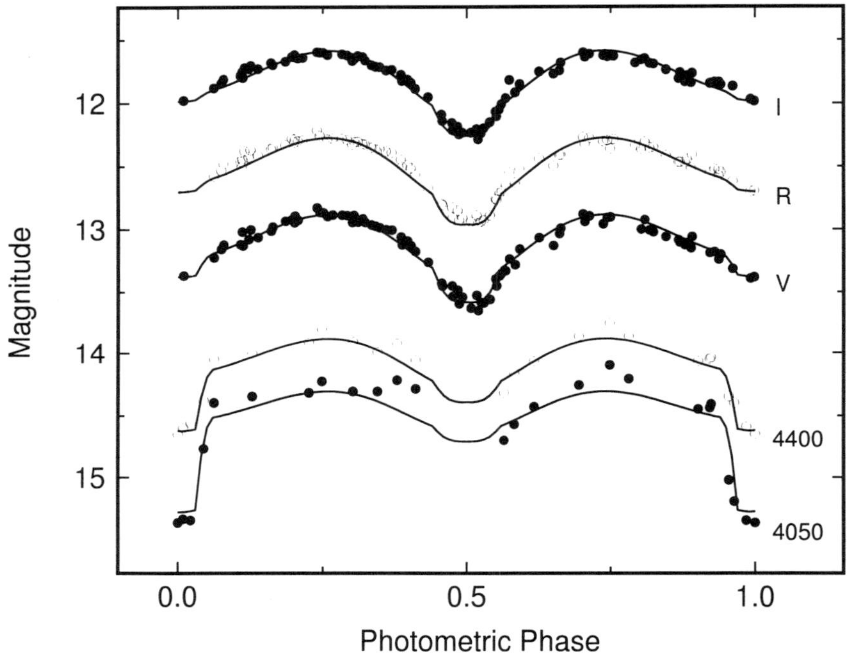

Fig. 5. Optical photometry of BG Gem from Benson et al. (2000) folded on the 91 day orbital period. The data are indicated by filled and open circles; model light curves from the Wilson-Devinney program are shown as solid lines.

Fig. 6 plots radial velocities of the K0 I secondary star and Hβ emission as a function of photometric phase. A circular orbit fits the velocities of the secondary star. The mass function is $M_1^3 \sin^3 i = (3.97 \pm 0.09)$ M$_\odot$ $(M_1 + M_2)^2$; $M_1 \approx 4$ M$_\odot$ $(1 + q)^2$ for sin i = 1. The Hβ velocities show the classical rotational disturbance produced by the eclipse of a rapidly rotating disk. The blue-shifted half of the disk is eclipsed first during ingress and is revealed first during egress. The apparent rotational velocity of $\sim$ 125 km s^{-1} implies disk material at $R_{H\beta} \approx$ 35 - 50 R$_\odot$(M_1/4 M$_\odot$). Together with the $R_{H\beta}$ derived from eclipses, this result confirms $q \sim 0.1$ and the orbital separation $A \approx 150$ R$_\odot$ derived from the light curve models.

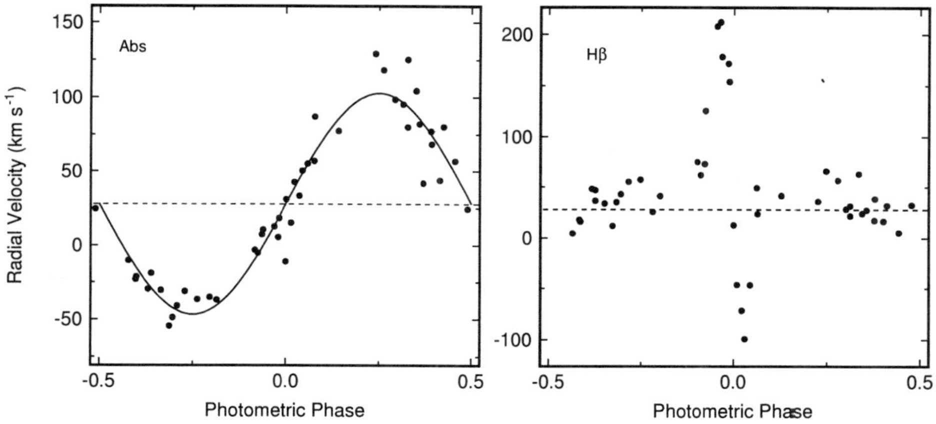

Fig. 6. Radial velocity curves for BG Gem. Data in the left panel show velocities for the K0 I absorption spectrum; the solid line is the best-fitting circular orbit to these data. Data in the right panel are velocities of the Hβ line. The constant velocity outside eclipse is indicative of a large mass ratio. The rotational disturbance at phase 0 is characteristic of the eclipse of a rotating disk.

The light curves and radial velocities demonstrate that the K0 supergiant (M_2 = 0.45 $M_\odot$) fills its Roche lobe and transfers material into an extended disk surrounding a massive primary star (M_1 = 4.5 $M_\odot$). The reddening is A_V = 1.65 $\pm$ 0.25; the distance is d = 2.25 $\pm$ 0.25 kpc. The disk contributes most of the blue light and has the flux distribution, $F_\lambda \propto \lambda^{-7/3}$ at 4000 - 6370 Å, expected for a steady-state disk. If the primary is a 4.5 $M_\odot$ B5 V star, the accretion rate would need to be a few $\times$ 10^{-6} $M_\odot$ yr^{-1}.

There is no evidence for optical radiation from a hot star at the center of the disk. A B5 V star should dominate the 4000 - 5500 Å light; we find no evidence for B-type absorption features on optical spectra (Fig. 4). An extended disk could occult the B star for $i \gtrsim 85°$, but it is difficult for a B5 star to produce enough photons to ionize H to the observed level. The extinction-corrected Hβ luminosity of 0.2 $L_\odot$ requires $\sim$ 4 $\times$ 10^{45} H-ionizing photons s^{-1}. This limit is one to two orders of magnitude larger than expected from a B5 V star. A boundary layer at the inner edge of the disk also produces too few ionizing photons, unless the accretion rate is much larger than few $\times$ 10^{-6} $M_\odot$ yr^{-1}.

Interestingly, a black hole is a good alternative to a main-sequence primary star. BG Gem has features in common with quiescent black hole binaries, including a) large ellipsoidal variations indicating extreme mass ratios (Shahbaz et al. 1994, 1996); b) modest emission line spectra, with strong, double-peaked H I lines and weaker, double-peaked He I lines (Orosz et al. 1998, McClintock et al. 1995) and c) disk-like energy distributions (McClintock et al. 1995). Other black hole binaries are more extreme X-ray sources than BG Gem (which has an average flux of -0.1 ± 1.0 mCrab at

2 - 12 keV from 100 days of Rossi XTE (ASM) data), but all other black hole binaries have orbital inclinations much smaller than the $i \gtrsim 80^\circ$ we derive for BG Gem (Chen et al. 1997). Thus, X-ray emission could be hidden behind the outer edge of the disk and BG Gem might contain a black hole primary.

To test the black hole hypothesis, we will acquire low resolution ultraviolet spectra of BG Gem with STIS in the next year. Ultraviolet spectra should allow us to distinguish clearly between a B star and a black hole. If BG Gem does contain a black hole, it would have the longest known period among black hole binary systems by a factor of 10. As the brightest black hole binary system, BG Gem would allow unique studies of the disk and secondary star impossible in other systems with current instrumentation. Finally, BG Gem would be the *first* black hole binary system identified without having undergone an X-ray nova outburst. Unknown selection effects are a major uncertainty in evolutionary models for black hole binaries. Knowledge of different formation channels is important for understanding how black hole binaries form and evolve and also places interesting constraints on the formation and evolution of different classes of progenitor binaries in the Galaxy (e.g., Kalogera 1999).

4. FINAL THOUGHTS

In the 100+ years since the birth of Cecilia Payne-Gaposchkin, stellar structure and evolution has gone from a fanciful dream to one of the most successful theories in astrophysics. Interacting binaries, including AG Peg and BG Gem, are a vital part of this theory. Detailed observational and theoretical studies demonstrate that the eruptions of AG Peg and other symbiotic novae result from thermonuclear runaways on the surface of a white dwarf. Although similar in many ways to the more well-studied outbursts of classical novae, the richness of symbiotic nova eruptions has expanded our understanding of thermonuclear runaways and improved our explanations of classical nova outbursts (Starrfield 1999). All classes of novae eject nuclear processed material into the Galaxy for incorporation into the next generation of stars. Quantifying the amount of recycled material gives us a better picture of the long term evolution of the Galaxy.

Other phenomena observed in symbiotic binary stars - such as coherent pulsations, flickering, jets and stellar winds - occur in a wide variety of binary and single stars. Synthesizing information derived from these systems has yielded important advances in mechanisms for mass accretion and mass ejection in active galactic nuclei and gamma ray burst sources in addition to binary stars (Livio 1999). Many of these studies have relied on observations from the Hubble Space Telescope and the Compton Gamma Ray Observatory. Future advances will include data from the Chandra X-ray Observatory and the Space Telescope Infrared Facility. These data provide us with the facts needed to test and hone our assumptions and our theories.

The 'new' interacting binary BG Gem provides a textbook example of analyzing modern observations with 'classical' techniques. By the early 1900's, deriving stellar parameters from eclipsing binaries - even those with distorted light curves - was a common, laborious practice with photographic photometry and spectroscopy. By the 1940s, 'peculiar' interacting binaries such as β Lyrae were well-known but poorly understood. Modern observations covering the electromagnetic spectrum coupled with advanced computer algorithms now yield important insights into these 'old' binaries. Satellite observations will eventually yield an answer to the puzzle of BG Gem, black hole or main-sequence star; a puzzle that probably would have amused and delighted C. Payne-Gaposchkin throughout her distinguished career.

ACKNOWLEDGMENTS

I thank M. Geller for a careful reading of the manuscript and acknowledge support from an archival grant from the Space Telescope Science Institute, AR-08369-01-A.

REFERENCES

Allen, D. A. 1980 MNRAS 190, 75
Baade, W. 1956 ApJ 123, 550
Belyakina, T. S. 1968 Astr. Zh. 45, 139
Belyakina, T. S. 1970 Astrofizika 6, 49
Benson, P., Dullighan, A., Bonanos, A., McLeod, K. K. & Kenyon, S. J. 2000 AJ 119, 890
Chen, W., Shrader, C. R. & Livio, M. 1997 ApJ 491, 312
Fleming, W. P. 1907 Ann. Harv. Coll. Obs. 47, 1
Gallagher, J. S. & Code, A. D. 1974 ApJ 189, 303
Gallagher, J. S., Holm, A. V., Anderson, C. M. & Webbink, R. F. 1979 ApJ 229, 994
Gallagher, J. S. & Starrfield, S. G. 1978 ARA&A 16, 171
Hutchings, J. B., Cowley, A. P. & Redman, R. O. 1975 ApJ 201, 404
Kalogera, V. 1999 ApJ 521, 723
Kenny, H. T., Taylor, A. R. & Seaquist, E. R. 1991 ApJ 366, 549
Kenyon, S. J. 1986 The Symbiotic Stars, Cambridge University Press
Kenyon, S. J., Mikołajewska, J., Mikołajewski, M., Polidan, R. S. & Slovak, M. H. 1993 AJ 105, 1573
Kenyon, S. J., Proga, D. & Keyes, C. D. 2001 AJ, submitted
Kenyon, S. J. & Truran, J. W. 1983 ApJ 273, 280
Kenyon, S. J. & Webbink, R. F. 1984 ApJ 279, 252
Livio, M. 1999 Physics Reports 311, 225
Lundmark, K. 1921 Astr. Nachr. 213, 93
McClintock, J. E., Horne, K. & Remillard, R. A. 1995 ApJ 442, 358
Merrill, P. W. 1916 Pub. Michigan Obs. 2, 71

Merrill, P. W. 1929a PASP 41, 255

Merrill, P. W. 1929b ApJ 69 330

Merrill, P. W. 1944 ApJ 99, 15

Merrill, P. W. 1959 ApJ 129, 44

Mürset, U., Nussbaumer, H., Schmid, H. M. & Vogel, M. 1991 A&A 248, 458

Orosz, J. A., Jain, R. K., Bailyn, C., McClintock, J. E. & Remillard, R. A. 1998 ApJ 499, 375

Payne-Gaposchkin, C. 1952 Ann. HCO 115, No. 22

Payne-Gaposchkin, C. 1957 The Galactic Novae, Amsterdam, North Holland

Proga, D., Kenyon, S. J., Raymond, J. & Mikołajewska, J. 1996 ApJ 471, 930

Proga, D., Kenyon, S. J. & Raymond, J. 1998 ApJ 501, 339

Shahbaz, T., Naylor, T. & Charles, P. A. 1994 MNRAS 268, 756

Shahbaz, T., van der Hooft, F., Charles, P. A., Casares, J. & van Paradijs, J. 1996 MNRAS 282, L47

Starrfield, S. 1999 Physics Reports 311, 371

Wilson, R. E. & Devinney, E. J. 1971 ApJ 166, 605

Wilson, R. E. 1994 PASP 106, 921

THE OLD GIANTS IN GLOBULAR CLUSTERS

Catherine A. Pilachowski

National Optical Astronomy Observatory, Tucson, Arizona[1]

ABSTRACT: The galactic globular clusters offer us insights into three important problems in astrophysics: the origin of the Milky Way, the formation of globular clusters, and the evolution of stars. The abundances of elements determined from the spectra of old giants in globular clusters reflect both the broad history of the Galaxy and the intricate physical processes occurring in star clusters and in individual stars as they evolve. Each globular cluster has its own story to tell.

1. THE HISTORICAL CONTEXT

At the time that Cecilia Payne-Gaposchkin arrived at Harvard, Shapley's great insights (Shapley 1919) about the globular clusters had established the size and structure of the Milky Way. It was not until the 1940's that Baade (1944) understood the division of Populations I and II, and not until the 1950's that Arp et al. (1952) detected the main sequences of the globular clusters, tying them to the emerging picture of stellar evolution and establishing their great age.

Spectroscopy of the integrated light of the globular clusters showed us that they are metal-poor compared to the Sun (Mayall 1946, Morgan 1956). Spectra of individual giants were analyzed by Kinman (1959) and by Helfer et al. (1959), giving us quantitative measures of the metallicities of globular clusters. Eggen, Lynden-Bell & Sandage (1962) brought these threads together into a tapestry of the formation of the Galaxy with its clusters. This model held until Payne-Gaposchkin's death in 1979.

The giants of globular clusters are the stars which tell us the most about the chemical composition of the clusters. Clusters are (mostly) metal poor, and the giants are both cool enough to show lines of many species (Payne 1925) and luminous enough that we can record their spectra with telescopes of moderate aperture. A flood of globular cluster spectroscopy began in the late 1970's with the

[1]Operated by the Association of Universities for Research in Astronomy, Inc. (AURA) under cooperative agreement with the National Science Foundation.

The Starry Universe: The Cecilia Payne-Gaposchkin Centenary
A. G. Davis Philip & Rebecca A. Koopmann, eds.
p. 135-145

C. A. Pilachowski
National Optical Astronomy Observatory
PO Box 26732
Tucson, AZ 85726-6732
CATYP@NOAO.EDU

commissioning of the new echelle spectrographs on 4-m class telescopes at KPNO, CTIO, ESO and the AAT. Since then, a few hundred cluster giants in more than 29 globular clusters have been observed at high spectral resolution.

Since 1979, the number of globular clusters with reliably determined metallicities (Harris 1996) has doubled. We now know that the metallicity distribution of the globular clusters is bimodal, with broad peaks near [Fe/H] = -1.6 and [Fe/H] = -0.6. [22] We understand that the Milky Way includes (at least) two populations of clusters, representing different epochs and perhaps mechanisms of formation.

2. CLUSTER-TO-CLUSTER VARIATIONS IN ABUNDANCE

Among the basic assumptions of the first half of the century was the certainty that stars shared a common composition. Mrs. Gaposchkin's thesis (1925) provided the first evidence that this was so, at least for elements heavier than hydrogen and helium. This truth held until the 1950's when Chamberlin & Aller (1951) reported on the metallicities of HD 19445 and HD 140283, finding them to be deficient in calcium and iron relative to hydrogen. Aller & Greenstein (1960) provided the first hints that even within the metals, the compositions of stars were not uniform, finding possible excesses of the α-elements later confirmed by Wallerstein (1962).

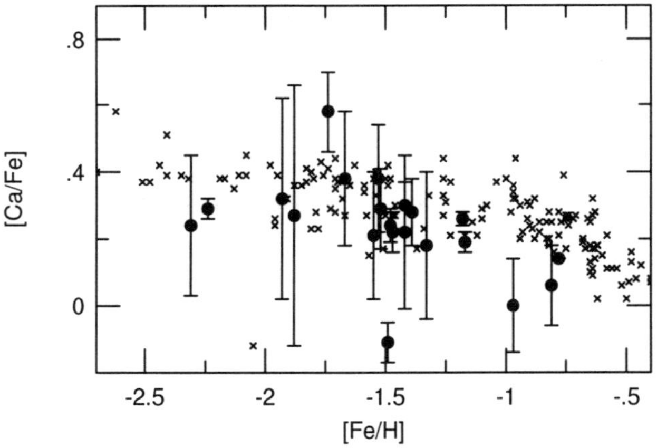

Fig. 1. [Ca/Fe] vs. [Fe/H] in globular clusters (filled symbols) and in halo field stars (crosses). The size of each error bar is set to the standard deviation for stars in that cluster.

The abundances of metals in globular clusters vary from cluster-to-cluster, not just in the scaled overall abundances, but in the detailed element-to-element abundance ratios. The variation of [Ca/Fe] with [Fe/H] is shown in Fig. 1 for both field halo stars (Fulbright 2000) and globular clusters. These

[2]I adopt the standard notation with $[A/B] \equiv \log_{10}(N_A/N_B)_{star} - \log_{10}(N_A/N_B)_{\odot}$, for elements A and B. Metallicity is equated here to the [Fe/H] value.

populations seem to behave similarly, following the same mean trend line with metallicity, and both populations include outlying points. The cluster Ruprecht 106 (Brown et al. 1997), for example, has a low [Ca/Fe] ratio in comparison to both the field halo stars and other clusters. Its low [Ca/Fe] ratio is explained as resulting from the formation of the cluster at a time after, rather than before, the Type Ia supernovae have raised the iron abundance (the excess of Ca in most clusters being produced by earlier Type II supernovae). Thus, the [Ca/Fe] ratio is a diagnostic of the time of cluster formation relative to the start of star formation (and element production) in a stellar system on timescales of 10^9 years.

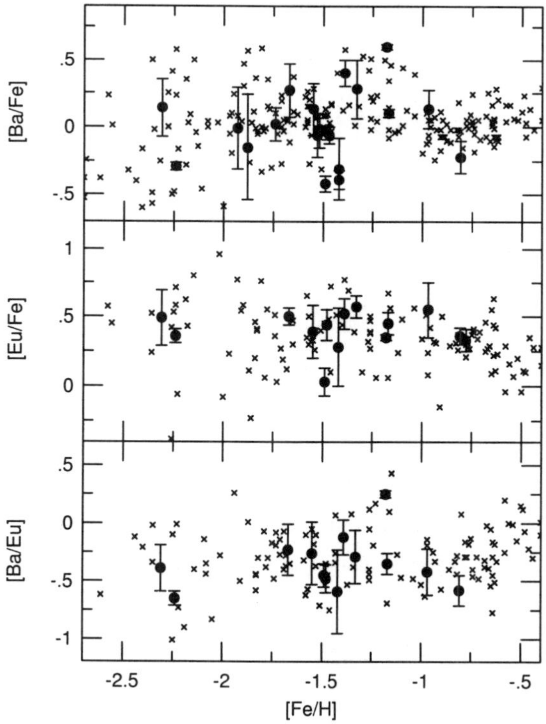

Fig. 2. Barium and europium abundance in globular clusters (filled symbols) and in halo field stars (crosses) vs. metallicity.

The behavior of [Ba/Fe] vs. [Fe/H] is shown in the upper panel of Fig. 2. The stellar data are taken from Fulbright (2000) and Burris et al. (2000). In this case, the clusters show a greater dispersion in [Ba/Fe] than does the field population. The dependence of [Eu/Fe] on [Fe/H] is shown in the middle panel. Here, both the field stars and the clusters show an excess compared to the Sun and a large scatter. In the lower panel of Fig. 2 is plotted the [Ba/Eu] ratio, which is a diagnostic of the relative contributions of the r-process and the s-process of nucleosynthesis (the solar ratio, with contributions from both is, of course, [Ba/Eu] = 0, while the pure r-process ratio is [Ba/Eu] = -0.89 (Burris et al. 2000)). The cluster M 4, at [Fe/H] = -1.2, has both a high [Ba/Fe] ratio and a high [Ba/Eu] ratio (Ivans et al. 1999), showing clear evidence of contributions of the s-process from asymptotic giant branch stars. The presence of s-process material in the globular clusters indicates that most cluster formation lagged the start of star/element formation by 10^8 years or so, the time scale for the evolution of the AGB stars which contribute s-process elements.

These cluster-to-cluster variations in the globular cluster mean abundances indicate qualitative differences in the timescales and scenarios for cluster formation. The clusters were not formed in a homogeneous fashion, but rather experienced different histories that are reflected in their current compositions.

3. STAR-TO-STAR VARIATIONS WITHIN CLUSTERS

Individual stars within globular clusters also differ one from another. These differences aren't restricted just to unusual clusters like ω Centauri, or just to light elements like carbon, nitrogen or oxygen, which are affected by the first dredge-up. Langer et al. (1998) reported that even in the classic metal-poor cluster M 92, not all of the stars have the same iron abundance. Based on high resolution, high signal-to-noise ratio spectra, one star, XI-19, is 0.17 dex more metal-poor that the other giants studied in the cluster. Metallicity variations within the cluster M 22 have also been suspected for some time. Brown & Wallerstein (1992) reported that M 22 giants with strong CN bands have roughly twice the metal content (Mg, Si, Ca and Fe) as stars with weak CN bands. More recently, the southern cluster NGC 3201 was found by Gonzalez & Wallerstein (1998) to have a similarly non-uniform composition. These variations in the iron abundance within clusters most likely result from self-enrichment by supernovae during the process of cluster formation. It is likely that as larger samples of cluster giants are studied with greater precision, more and more such examples will be found.

Star-to-star variations in the strengths of the CN and CH bands (and in the carbon and nitrogen abundances) are found in the giants of most globular clusters. Often, the CN and CH distributions are bimodal among the giants. In the relatively nearby and metal-rich cluster M 71, CN and CH strengths have been surveyed both in giants (Penney et al. 1992) and in stars near the main-sequence (Cohen 1999) turnoff. Data from both studies are plotted in Fig. 3, which illustrates both that CN and CH are anti-correlated and that the stars fall into two distinct groups that are preserved as the stars evolve. As the M 71 stars evolve from the main sequence to the lower giant branch, the distributions of both CN and CH become increasingly bimodal, presumably due to the first dredge-up, and possibly due to additional mixing as the stars evolve up the giant branch.

The decline in the carbon abundance for stars on the giant branch is illustrated in Fig. 4, from

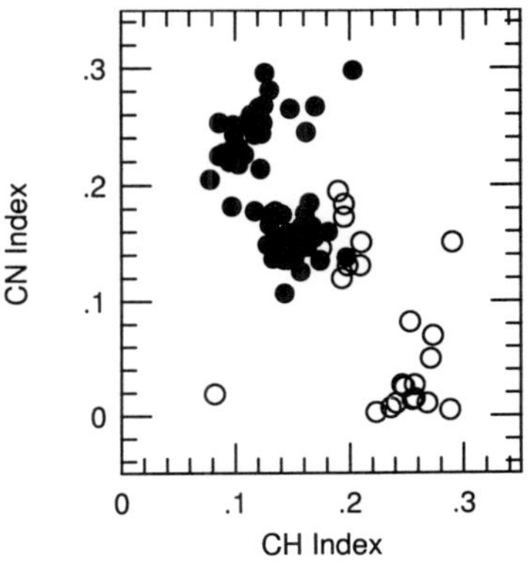

Fig. 3. CH vs. CN indices for main-sequence stars (Cohen 1999, filled circles) and low luminosity giants (Penney et al. 1992, open circles) in M 71. Note that the indices are defined differently in the two samples, so that the giants and dwarfs cannot be compared directly in this plot.

Bellman et al. (2001), for the globular cluster M 92. The carbon abundance continues to drop all along the upper giant branch, at luminosities well above the first dredge-up. Clearly, the convective envelopes of the cluster giants are exposed to nuclear processing which alters their abundances.

Star-to-star abundance variations are not restricted just to carbon and nitrogen. Fig. 5, taken from Ivans et al. (1999), shows the remarkable oxygen vs. sodium anti-correlation in several globular clusters. Stars with oxygen deficiencies are all sodium rich, suggesting that the process which depletes oxygen also enhances the sodium abundance. Changes in the sodium abundance with evolution along the giant branch in M 15 were found by Pilachowski et al. (1997), and confirmed by Cavallo & Nagar (2000), who found a marked rise in both the sodium and the aluminum abundance just at the tip of the giant branch in M 13.

These effects on oxygen, sodium, and aluminum, as well as on carbon and nitrogen, can be explained by proton-capture nucleosynthesis on the giant branch (Denisenkov & Denisenkova 1990, Langer et al. 1993, Cavallo et al. 1996). In addition to the traditional CN cycle associated with

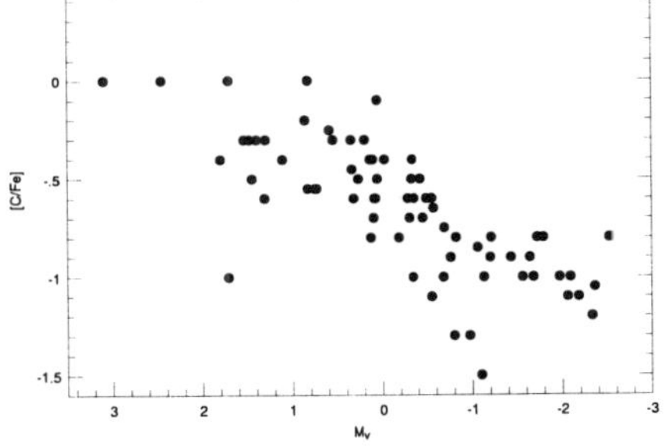

Fig. 4. Carbon abundances in M 92 giants from Bellman et al. (2001)

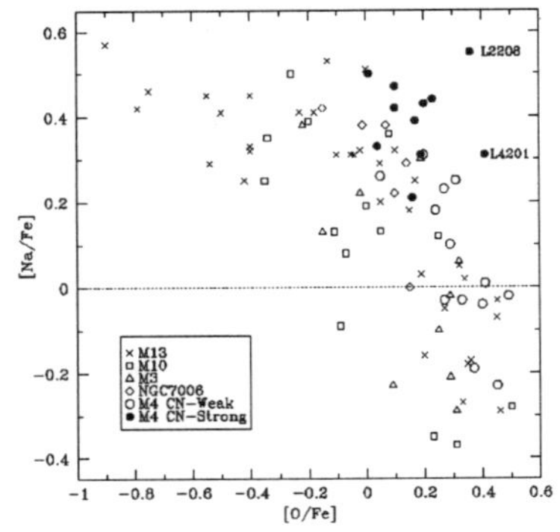

Fig. 5. [Na/Fe] vs. [O/Fe] for stars in several globular clusters from Ivans et al. (1999).

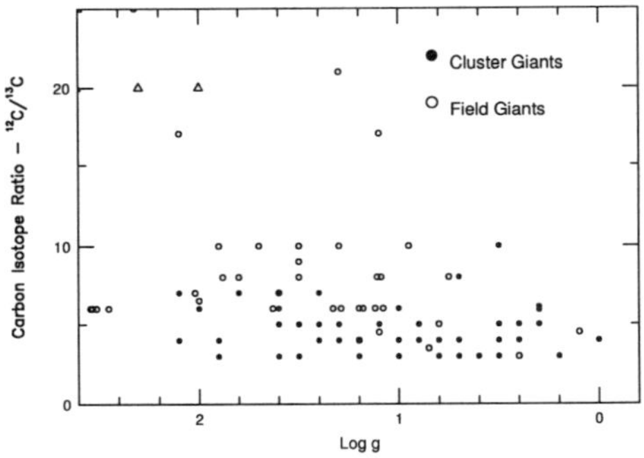

Fig. 6. Carbon isotope ratio vs. log g for field and cluster giants. The data have been assembled from a variety of literature sources. The isotope ratios in cluster giants are closer to the CNO equilibrium value than are the field giants.

the hydrogen burning shell in low mass giants, the ON, NeNa and MgAl cycles also appear to be operating, and deep mixing must connect the convective envelope to a depth where hydrogen burning occurs. The NeNa cycle results in the conversion of abundant neon to less abundant sodium, while the MgAl cycle converts magnesium into aluminum. Deep mixing may begin at the luminosity function bump on the lower giant branch, and continue up to the tip, driving more and more severe changes to the abundance mix of the light elements. All elements at least through aluminum are affected, although the fraction of stars affected by deep mixing varies from cluster to cluster. Rotation may be implicated in driving the deep mixing. Associated with changes to the abundances of C, N, O, Ne, Na, Mg and Al may be an increase in the abundance of helium.

The helium abundance is a factor in the subsequent stellar evolution on the horizontal branch and beyond, and the extent of deep mixing may play a role in determining horizontal-branch morphology. That globular clusters with similar metallicities may display different horizontal-branch morphology is known as the second parameter problem. Age differences must certainly play a role, and other factors (helium abundance, CNO abundance) may also be significant. Deep mixing and proton-capture nucleosynthesis, through the effect on the helium abundance, may well be one of the factors affecting the color of stars on the horizontal branch in some clusters.

Globular cluster giants and field giants appear to differ in their susceptibility to deep mixing and proton-capture nucleosynthesis. The effects seen so often in the cluster giants do not appear in isolated field stars. For example, the $^{12}C/^{13}C$ ratios are lower in cluster giants than in field stars of the same luminosity and metallicity, as shown in Fig. 6. With increased mixing the isotope ratio drops from its first dredge-up value to approach the equilibrium value from CN cycle burning. The lower isotope ratios in cluster giants compared to field giants suggest more extensive mixing.

Similarly, the magnesium abundances in cluster giants reflect more extensive NeNa- and MgAl-cycle processing. The field population contains few or no giants with high sodium and low magnesium abundances, while such stars are relatively common among cluster giants, as shown in Fig. 7. The cause of these differences between the field and cluster populations is not clear, but it may be associated with differences in initial rotation or in subtle tidal effects in the dense stellar environment of globular clusters. Since the halo field must contain refugee stars from globular clusters, it would seem that rotation alone cannot be the answer.

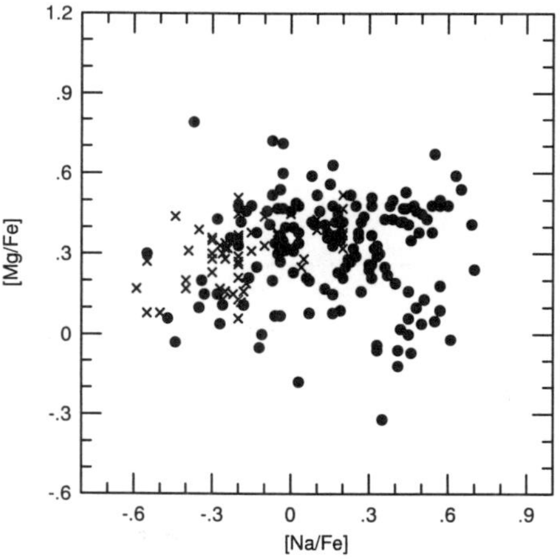

Fig. 7. [Mg/Fe] vs. [Na/Fe] for field halo giants (crosses) and globular cluster giants (filled circles).

Additional evidence for mixing in globular cluster giants may also be found in their lithium abundances. Chris Sneden (this volume) has already reviewed what is known about lithium in field halo giants. The recent discovery of a lithium-rich giant in the globular cluster M 3 (Kraft et al. 1999) has established that lithium giants also occur in globular clusters (lithium is not normally detected in the spectra of halo giants because of the low initial abundance of lithium, convective dilution, and subsequent giant-branch depletion). A survey of more than 700 cluster giants with the Hydra fiber spectrograph on the 3.5-m WIYN telescope suggests that lithium-rich giants are as common among the cluster population as among field halo giants (a few percent). Lithium-rich giants are found all along the cluster giant branches, but seem to concentrate predominantly at lower luminosities just above the luminosity function bump. While the origin of the lithium is still not clear, it may be associated with a mixing episode that occurs at or above the point where the hydrogen burning shell passes the composition discontinuity left from the deep penetration of the convective envelope on the lower giant branch. Is it possible that such low mass giants serve as a source for galactic lithium?

While the evidence points to deep mixing and proton-capture nucleosynthesis on the giant branch to explain many of the abundance variations observed in globular cluster giants, intrinsic star-to-star differences also occur. As noted above, some clusters show a dispersion in the iron abundance, which can only be attributed to processes like self-enrichment from early Type II supernovae. The iron abundance is not the only element affected, of course, as Fig. 8 from Briley et al. (1996) shows.

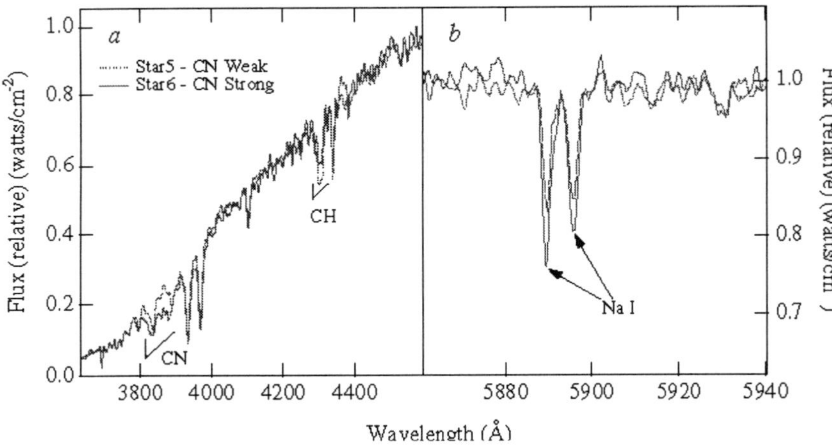

Fig. 8. Briley et al. (1996) found anti-correlated variations in the CH and CN and sodium line strengths in 47 Tucanae main sequence stars.

4. THE SPECIAL CASE OF ω CENTAURI

No discussion of abundance variations in globular clusters would be complete without mention of ω Centauri, the Milky Way's most massive star cluster. ω Centauri is well known for containing stars with a range of metallicity. Pancino et al. (2000) have recently reported a new color-magnitude diagram for the giants in ω Cen which shows perhaps as many as four distinct metallicity sub-groups. The cluster population is dominated by metal-poor stars with [Fe/H] $\sim$ -1.7, but it also includes a significant population of stars near [Fe/H] $\sim$ -1.3, and smaller populations at metallicities near [Fe/H] $\sim$ -0.7 and perhaps even near solar metallicity. None of these sub-populations seems to be mono-metallic, but each seems to have some breadth in metallicity. The most metal rich stars are distributed differently on the sky than the metal-poor stars, and may represent a merged remnant rather than stars originally formed as part of the Milky Way.

Synthesizing several recent studies of the composition of stars in the cluster, Norris (2000) suggested that the time of self-enrichment in ω Cen spread over two to three Gyr, and that contributions from both Type II supernovae and AGB stars can be seen in successive generations of stars. Evolutionary mixing in the giants also complicates the interpretation of the C, N, O, Na, Mg and Al abundances, as it does in other clusters. The apparent merger of sub-systems represented by the metal-rich stars and the time scale for self-enrichment both suggest that ω Cen formed in a low-density region outside the Milky Way, and may itself have been accreted by the Galaxy in a merger event.

Smecker-Hane et al.'s (2000) recent color-magnitude diagram of the Carina dwarf spheroidal galaxy also shows the effects of extended or episodic star formation, albeit with a different history. In the case of Carina, star formation seems to have occurred in bursts over a longer period of time than in ω Cen, producing multiple main sequence turn-offs and sub-giant branches merging onto overlapping giant branches. These two examples, ω Cen and Carina, together illustrate the complexity that may result in the color-magnitude diagrams and chemical abundances of globular clusters from the detailed history of each cluster's origin and star formation history.

5. SUMMARIZING OUR KNOWLEDGE OF GLOBULAR CLUSTER COMPOSITIONS

The globular clusters are the simplest stellar systems we have for studying chemical evolution, yet the compositions of present-day clusters reflect a colorful palette of phenomena. The star formation history of the original gas, and of the proto-globular clusters themselves contributed to the chemical mix, perhaps in different ways in each cluster. Mergers of sub-systems with different histories may also complicate the chemical evolution of globular clusters, as we seem to see in ω Cen. The galactic globular clusters show diverse origins. Some formed in the halo, some were captured from other systems or were accreted by the Milky Way through mergers with other galaxies, and some were clearly formed in the proto-disk of the Galaxy. Interestingly, none seemed to have formed recently, although there is evidence of young populations of globular clusters in other Local Group galaxies, as described in this volume by Paul Hodge. We know from studies of external galaxies that globular clusters form in many environments: halos, disks, bulges and elliptical galaxies, and in tidal tails. Most or all of these formation environments appear to be represented within the family of galactic globular clusters. The globular clusters tell us that at the time of cluster formation, the Milky Way was as chaotic as the distant galaxies seen in the Hubble Deep Field.

The compositions of stars in globular clusters are also complicated by nucleosynthesis and deep mixing, which rearranges the abundances of the light elements ($Z<14$). Proton-capture nucleosynthesis is not limited to the traditional CN, or even CNO, cycle, but includes nucleosynthesis via the NeNa- and MgAl-cycles as well. The interpretation of the abundances of these elements must include an understanding of how these processes contribute. Oddly, these processes do not seem to affect the field halo population at the same metallicities and luminosities as the cluster giants. Why globular cluster giants are more susceptible to deep mixing is a question that still must be answered.

In her autobiography, Cecilia Payne-Gaposchkin (Payne-Gaposchkin & Haramundanis 1984) wrote "...it is well for us that knowledge comes to us gradually. If we were faced with the full complexity of the facts at the beginning of our search, we should be so bewildered that we might give up in despair." Surely this applies as well to the globular cluster giants, whose complex chemistry was only hinted in her lifetime. How she would have enjoyed this unfolding story!

ACKNOWLEDGMENTS

I thank the Harvard College Observatory and the Smithsonian Astrophysical Observatory for supporting this Centenary celebration of the life and contributions of Cecilia Payne-Gaposchkin, and the organizing Committee Owen Gingerich, Andrea Dupree, Katherine Haramundanis, Dave Latham, A. G. Davis Philip and Virginia Trimble for their inspiration and hard work in organizing this meeting.

REFERENCES

Arp, H. C., Baum, W. A. & Sandage, A. R. 1952 AJ 57, 4

Aller, L. H. & Greenstein, J. L. 1960 ApJS 5, 139

Baade, W. 1944 ApJ 100, 137

Bellman, S., Briley, M. M., Smith, G. H. & Claver, C. F. 2001 PASP (submitted)

Briley, M. M., Smith, V. V., Suntzeff, N. B., Lambert, D. L., Bell, R. A. & Hesser, J. E. 1996 Nature 383, 604.

Brown, J. A. & Wallerstein, G. 1992 AJ 104, 1818

Brown, J. A., Wallerstein, G. & Zucker, D. 1997 AJ 114, 180

Burris, D. L., Pilachowski, C. A, Armandroff, T. E., Sneden, C., Cowan, J. J. & Roe, H. 2000 ApJ 544, 302

Cavallo, R. M. & Nagar, N. M. 2000 AJ 120, 1364

Cavallo, R. M., Sweigart, A. V. & Bell, R. A. 1996 ApJ 464, L79

Chamberlain, J. W. & Aller, L. H. 1951 ApJ 114, 52

Cohen, J. G. 1999 AJ 117, 2434

Denisenkov, P. A. & Denisenkova, S. N. 1990 Sov. Astron. Lett. 16, 275

Eggen, O. J., Lynden-Bell, D. & Sandage, A. R. 1962 ApJ 136, 748

Fulbright, J. P. 2000 AJ 120, 1841

Gonzalez, G. & Wallerstein, G. 1998 AJ 116, 765

Harris, W. E. 1996 AJ 112, 1487 (the 1999 June 2 on-line version of this catalog was used)

Helfer, L. H., Wallerstein, G. & Greenstein, J. L. 1959 ApJ 129, 700

Ivans, I. I., Sneden, C., Kraft, R. P., Suntzeff, N. B., Smith, V. V., Langer, G. E. & Fulbright, J. P.

1999 AJ 118, 1273

Kinman, T. D. 1959 MNRAS 119, 538

Kraft, R. P., Peterson, R. C., Guhathakurta, P., Sneden, C., Fulbright, J. P. & Langer, G. E. 1999 ApJ 518, L53

Langer, G. E., Fischer, D., Sneden, C. & Bolte, M. 1998 AJ 115, 685

Langer, G. E. & Hoffman, R. & Sneden, C. 1993 PASP 105, 301

Mayall, N. U. 1946 ApJ 104, 290

Morgan, W. W. 1956 PASP 68, 509

Norris, J. E. 2000 Mixing and Diffusion in Stars: Theoretical Predictions and Observational Constraints, 24th meeting of the IAU, Joint Discussion 5, August 2000, Manchester, England

Pancino, E., Ferraro, F. R., Bellazzini, M., Piotto, G. & Zocalli, M. 2000 ApJ 534, 83

Payne, C. H. 1925 Harvard Coll. Obs. Monogr. 1

Payne-Gaposchkin, C. & Haramundanis, K. 1984 Cecilia Payne-Gaposchkin: an autobiography and other recollections, Cambridge University Press, Cambridge

Penney, A. J., Smith, G. H. & Churchill, C. W. 1992 MNRAS 257, 89

Pilachowski, C. A., Sneden, C., Kraft, R. P. & Langer, G. E. 1997 AJ 112, 545

Shapley, H. 1919 ApJ 49, 311

Smecker-Hane, T., Stetson, P. & Hesser, J. 2000 (in preparation)

Wallerstein, G. 1962 ApJS 6, 407

OPEN CLUSTERS: PROBING STELLAR AND GALACTIC EVOLUTION

Kenneth Janes

Boston University, Boston, Massachusetts

ABSTRACT: The open clusters are the nearest and most familiar of stellar systems. Indeed, the stars in open clusters are representative of typical galactic stars in every respect. Some 1200 open clusters are known ranging in age from the still-forming Orion cluster to systems such as NGC 6791 that are of the order of eight billion years in age. Some open clusters have as few as a dozen members, while the most populous have somewhat more than 10,000 members.

1. HISTORICAL CONTEXT

Four key events over the past century illustrate the scientific importance of the open clusters. In 1908 Ejnar Hertzsprung constructed graphs of the magnitudes vs. colors of stars in several clusters in an attempt to find a way to separate cluster members from non-members. His version of what is now known as the Hertzsprung-Russell diagram is still a standard tool for deriving the physical properties of stars, including in particular their ages and life histories.

Some years later, Robert Trumpler (1930, PASP, 42, 214) in effect discovered the interstellar medium by looking at open clusters. He found that the more distant clusters were fainter than would be expected from their angular sizes, and were redder than the more nearby ones. Thus he found that the galactic disk is filled with a thin absorbing material.

Cecilia Payne Gaposchkin in her text "Introduction to Astronomy" (1954, New York: Prentice-Hall) wrote (p. 469), "The problem of stellar evolution has never been satisfactorily solved". But within a few years, Martin Schwarzschild, Fred Hoyle, Icko Iben, Allan Sandage and others had worked out the modern theory of stellar structure and evolution that explained in great detail the appearance of the Hertzsprung-Russell diagrams of star clusters. It is now possible to derive at least the relative ages of star clusters with considerable precision.

The final "event" of this list actually extends across the entire past century - the use of clusters to define the cosmic distance scale. Harlow Shapley's work with the globular clusters that placed the galactic center thousands of parsecs away in the direction of Sagittarius set the stage for progressively more precise calibrations of the distance scale. By 1998, M. A. C. Perryman, et al. (A&A, 331, 81) were able to find the distance to the Hyades star cluster (45 parsecs away) with an uncertainty of only $\pm$ 1 parsec using the Hipparcos satellite. When the globular clusters are tied into this result, it will be finally possible

The Starry Universe: The Cecilia Payne-Gaposchkin Centenary
A. G. Davis Philip & Rebecca A. Koopmann, eds.
p. 147-148

Kenneth Janes
Dept. of Astronomy
Boston University
725 Commonwealth Ave., Boston, Mass. 02215
JANES@HYADES.BU.EDU

to derive the absolute ages of the oldest stars to better than 10 %.

2. SO WHY DO PEOPLE STILL LOOK AT CLUSTERS?

The easiest way to collect a homogeneous sample of stars, with known (or at least knowable) properties is to select stars in a cluster. a) Cluster stars all have (nearly) the same age. b) They are all at (nearly) the same distance. c) They all have (nearly) the same chemical composition. d) These properties can be measured rather accurately. E) Dynamically, clusters are simple systems of point masses.

3. SOME CURRENT ISSUES

First is the matter of cluster formation and evolution. Do all stars form in clusters? What is the initial mass function and is there mass segregation in the cluster as it is forming? What is the dynamical evolution of clusters?

The lowest luminosity stars are of interest for a variety of reasons: White dwarf stars can be used to derive the ages of the oldest clusters; there are searches underway for brown dwarf stars in the youngest and nearest clusters such as Orion, the Pleiades and the Hyades; and because so many stars can be imaged at one time in a cluster, it may even be possible to detect rare planetary transits.

Clusters provide a rich field for investigating stellar activity. Satellite X-ray images of the Hyades can uncover stars with active chromospheres. Spectra of both giants and dwarfs in the old cluster M 67 (a cluster virtually the same age as the Sun) can be used to explore the range of activity levels in solar-type stars. Other clusters, both younger and older can be searched for signs of chromospheric and photospheric activity (starspots), to learn how activity declines with the age of a star.

Finally, clusters are still important to help address a number of questions of stellar and galactic evolution. What are the hot SdO/SdB stars seen in some clusters? What are the blue straggler stars? What is the relation between lithium abundance and the age and main-sequence location of a star? What is the age of the galactic disk?

ACKNOWLEDGMENT

From 1960 to 1963, I worked in Professor Payne-Gaposchkin's office helping with her massive project to catalog the variable stars in the Small Magellanic Cloud. I was so much in awe of her brilliance that it never occured to me at the time how unique it was for a young man to have a woman as his scientific mentor. I certainly owe my own modest scientific success to her help and encouragement

CECILIA PAYNE-GAPOSCHKIN AND THE REST OF THE UNIVERSE

Virginia Trimble

University of Maryland & University of California, Irvine

ABSTRACT: A brief overview is given of Cecelia Payne-Gaposchkin's work on supernovae, star formation, and galactic structure and evolution, including connections to earlier and later work on these topics. A few side-light subjects with which she has been more often identified are also presented

1. INTRODUCTION

Only after providing the above title to the organizers did I realize that it incorporates a certain ambiguity. It sounds as if "rest of ...." might mean all the the universe that was not CHPG, too much territory even for someone whose home address ranges across more than 7000 ZIP codes. It was intended to mean her work on topics that would not be discussed by other speakers at this symposium, simply because her interests were extraordinarily wide ranging and a 1.5 day meeting has to be somewhat focussed. She would surely have said it better!

In the course of the meeting, participants heard the honoree called "Payne" (by her daughter), "Mrs. G" (by several former Harvard students), "Cecelia Payne-Gaposchkin" (by speakers who had never met her), "Cecelia" by her few remaining approximate contemporaries, and probably other things. Although I knew her slightly in the last, roughly, decade of her life and spoke with her on a number of occasions, I believe I always avoided calling her by any specific name. This is a personal idiosyncrasy of long standing. (I called my parents "you guys" and in general arrange to be within the field of view before addressing anybody, thereby avoiding the need for any recognition signal, except, perhaps, "just the person I was hoping to see here".) The two letters that I have on file are addressed "Dear Virginia" and signed "Cecilia", but she will appear in this chapter as CHPG, both in declarative sentences and in references. The "H" stood for Helena (not hydrogen, though this would have been appropriate) and may or may not have made any contribution to her life-long interest in classics.

The topics addressed below are supernovae, star formation, galactic structure and evolution, and "the rest of the rest".

The Starry Universe: The Cecilia Payne-Gaposchkin Centenary
A. G. Davis Philip & Rebecca A. Koopmann, eds.
p. 149-158

Virginia Trimble
Physics Dept. University of California
Irvine, California 92697-4575
VTRIMBLE@UCI.EDU

2. SUPERNOVAE: THE OUTSIDES

Most supernova research of the years (once SNE were accepted as a significant class of astrophysical event) can be divided in two classes: surface phenomena (light curves, spectra, statistics and their interpretation) and interior processes (energy sources and connection with the rest of astrophysics, shock propagation, etc.). CHPG worked directly on the former and expressed opinions on the latter in her books.

The first two supernova spectra to be described, S And (1885) and Z Cen (1895) and roughly the next ten were all of type I. Everyone agreed that there was always a continuum, which declined after maximum light, and most of the early workers described the fluctuations of intensity with wavelength as very broad bright line or emission features, by analogy with the spectra of novae. This includes CHPG (1936a on Z Cen, 1936b on S And). She also addressed the issue of photospheric temperature. Baade & Zwicky (1934) had believed this to be in the range 10^{5-6} K, having assumed a photospheric size of 10^{13} cm, like those of common novae, and deduced that the then undetectable ultraviolet and X-ray radiation would carry away most of the copious energy they had provided by making supernovae out of the collapses to neutron stars.

CHPG, in contrast, concluded that the overall similarity of the spectra to those of novae implied a comparable temperature of 10 - 20,000 K or less and so (with the distance scale then in use) "only" about 10^{48} ergs in electromagnetic radiation, compared to the $10^{52\pm1}$ deduced by Baade and Zwicky. (This rises to 10^{49} ergs or more with a modern extragalactic distance scale.) But she also pointed out (not uniquely) that the spectral features were much broader than those of common novae, perhaps a total width of 12,000 km/sec, vs. about 1000. Zwicky (1936a, b) then fired back the rebuttal that this sort of velocity, imparted to at least one solar mass (needed to keep the ejecta optically thick for a year or more) took you back up at least to 10^{51} ergs. And he was, of course, right. The rest of the available 10^{53} ergs he put into non-visible radiation, cosmic rays and energy somehow stored in the neutron star. (One rotating at break up would, of course, need all the available energy.)

It was against this background that CHPG & Fred Whipple [(1940), Whipple & CHPG (1941)] attempted to synthesize surpernvova spectra, with particular reference to that of 1937c, which had been followed for more than a year; and, yes, they used the word "synthesize". They took known transitions of common elements in well-studied ionization levels appropriate to temperatures of 5,000,000 - 20,000 K (H I, He I & II, C II, III & V, N II to V, O II to VI, Na I, Fe II & III), assigned f values to the various transitions, using a combination of sum rules, calculated Einstein coefficients, and laboratory data, broadened the emission lines to (fullwidth at zero intensity) 12,000 km/sec, added up the ones that should be present at different particular temperatures, and adjusted the elemental composition until they got the best fit to spectra observed at different times through the event.

Important results included a) the fits looked pretty good, b) zero hydrogen was fine, c) iron was enhanced and d) so were He, carbon and nitrogen (which they associated with Bethe's CN cycle).

Item (b) is definitely true and can be credited to them; (c) is also true though possibly not established by their syntheses; and (d) is not necessarily true. Both then went on to other things. Of Whipple's many contributions I will mention here only the very prescient 1942 talk on star formation, not published until four years later (Whipple 1946).

And there the problem of the Type I supernova spectra sat for roughly 35 years, through the recognition that there are also Type II's dominated by hydrogen features (Minkowski 1940). From time to time, other spectroscopists took a shot at it. The approach of Mustel (1970) was particularly interesting. His inventory of elements and transitions was, of course, more complete and accurate that that of CHPG and Whipple. But, in addition, he assumed that the spectral features were broad absorption rather than broad emission. Strangely, his fits were also "pretty good", though not widely believed. Indeed, I heard him give a colloquium on the subject (in Cambridge, England) in about 1970, and thought silently, "he's full of bananas; if it were that simple the problem would have been solved long ago".

I do not know whether CHPG ever met E. R. Mustel. She never collaborated with him and was presumably not encouraged at home to think highly of Soviet astronomers after 1933. Her one Russian co-author (besides her husband) was Gerasimovich, on a single 1932 paper on SS Cygni stars.) But, between them, CHPG and Mustel had the answer. For the spectra of Type I supernovae can indeed be fit by a superposition of strong transitions of common ions (excluding hydrogen and, sometimes, helium), but each feature must be assigned both an emission and an absorption part, that is, a P Cygni profile. Modern understanding along these lines came soon after Mustel's efforts, (Kirshner et al. 1973, Branch & Patchett 1973). There was some communication between the two groups, but only Kirshner et al. cited Whipple & CPHG. They confirmed the absence of hydrogen and the conspicuousness of iron features, and we will follow this trail no further.

CHPG never published on the spectra of Type II events, but she correctly states in "The Galactic Novae" (CHPG 1957) that they are hydrogen-rich and considerably commoner than the Type I's.

3. SUPERNOVAE: THE INSIDES

CHPG expressed views on the nature and energy sources of supernovae formally in the variable stars book (CHPG & Gaposchkin 1938) and in "Galactic Novae" (CHPG 1957) and less formally in her introductory text (CHPG 1954). In 1938 supernovae unquestionably exist, with credit to papers around 1920 by Shapley (who then at least thought the whole idea rediculous) and Lundmark, but not to contemporary work by Curtis or to Baade and Zwicky. The Gaposchkins regard the name as "somewhat illogical", but not worse than lots of other astronomical terminology, and anyhow too well established to do much about. The authors concluded that supernovae form a continuum with the common novae and share the same physical mechanism, but occurring perhaps in giant rather than dwarf stars.

Their list of events included Tycho's and Kepler's new stars, Z Cen and S And, and a handful of early 20th century discoveries, but none of what Lundmark (1935 publication date on the volume but 1932 for his paper and so the first printed use of the word supernova) had called the "pre-Tychonic nova", not even 1054. They expressed the view that at most one of his ten or so events was likely to have been a supernova, on the grounds that such an event anywhere in the Galaxy would reach apparent magnitude of at least zero. All of Lundmark's candidates did, and interstellar absorption seems to have been completely forgotten, despite Trumpler's 1930 demonstration of 1 mag/kpc of extinction in the galactic plane.

By 1954, the Crab Nebula has entered the inventory, in fact stars in it, and she discussed three potential energy sources. These are a) stellar collisions (put forward by Whipple in 1939 but dismissed by CHPG on the grounds that it would predict many more faint events than bright ones), b) "sudden release of internal energy" which is declared appropriate for novae, and c) "sudden radical internal transformation", very probably from red giant to a white dwarf, declared appropriate for supernovae. This last is largely the fault of Chandrasekhar, who put forward the idea at a 1939 symposium in Paris, stuck by it for at least two decades, and somehow persuaded many distinguished astronomers, including Baum, Spitzer and Schwarzshild, down almost to the time of the discovery of pulsars (e.g., Abell 1964). See Trimble (2001) for more on this point.

The 1957 discussion is an enormously valuable guide to the earlier literature, but leaves one in just a little doubt about how long it had been since CHPG herself had read some of the papers. Supernovae are now described as not being part of a continuum with novae (or U Gem stars, now dwarf novae), and Baade & Zwicky (1934) are duly credited with collapse to a neutron core as the energy source. But she comes out in favor of two different sorts of nuclear processes for the two types. These are fusion up to Cr^{254} for SNI's [credited to Baade et al. (1956); the clearer and much more frequently cited B^2FH (Burbidge et al. 1957) had not yet quite appeared as CHPG was writing], and "capture of protons with release of gamma rays" for SN II's. Credit for the latter, which she also calls "thermal blow-up" is assigned to Schatzman (1946). What he actually wrote in that paper, called theory of supernovae, was "L'energie liberee au cours de l'exposion est due a la transformation de l'energie cinetique d'exlusion des electrons en energie d'agitation thermique". That is, he had in mind clawing back the energy trapped in the necessarily large velocities of degenerate electrons and using it to power an explosion. I do not see any physically possible way to do this, and, even if it were possible, it is certainly not explosive nuclear fusion.

Also cited, but not given credit for being first with anything, are Gurevich & Lebedinsky (1946), whom I came across independently and quite by chance while prowling among old journals in a Maryland basement. (My original goal had been to find a paper by Gamow from before he left the USSR - I failed but also got distracted.) The discussion in Gurevich & Lebedinsky is very brief and and has no real calculations, but they do speak of "degenerated" nuclear explosions, which, if triggered on a stellar surface will make a nova with the star surviving and, if triggered centrally, will make a supernova, with the star destroyed. Hard to say it better in so few words! Incidentally, most reviews (including, I fear, some of mine) credit the nuclear explosion picture to Hoyle & Fowler (1960) with Schatzman & Mestel relegated to a footnote at best, and Gurevich & Lebidinsky missing completely.

Finally, it is irresistible to quote from p. 274 of CHPG (1957) "Type I supernovae belong to Population II, Type II supernovae to Population I". We have probably all said it, though not always so succinctly, and I wonder whether she was first to phrase it as a sort of epigram.

4. THE STORY OF STAR FORMATION

A few years ago, the custom of beginning the annual Goddard/Maryland October workshop with a historical introduction by Trimble required that I find out about star formation. My first step was, of course, to ask Owen Gingerich. His response, "there is no history of star formation", was initially a bit unsettling. But, it turned out that he did not mean that the subject had no intellectual underpinnings, only that nobody seemed to have written about it. Trimble (1997), which rejoiced in the dubious title "The Fourth Day of Creation" is, therefore, my primary reference for saying that to title a book "Stars in the Making" in 1952 (CHPG 1952) was actually a pretty brave thing to do.

What had happened was roughly the following. The expansion time for the universe implied by Hubble's (1929 and many later publications) own value of his constant came very close to the age of the Earth as measured from U and Th decaying to lead in rocks. Then Bethe's (1939) CN cycle as the source of stellar energy yielded a very similar time scale for the Sun and stars (whose interior initial hydrogen abundance was not so large then as is now). The step from "most things are about the same age" to "everything is the same age" was all too easy to take. Thus technical, text and popular books were saying things like this:

a) "Conditions in the universe were so crowded and presumably so different from those which now prevail ... It is tempting to look upon this as a time of turbulence, in which various things happened which are very difficult to account for otherwise – such as the origin of the solar system and of double stars." (1938 edition of Russell, Dugan & Stewart).

b) "The theory of the Expanding Universe indicates that a catastrophe took place 3×10^9 years ago and it is tempting to place the origin of stars and stellar systems tentatively at the epoch of this catastrophe." (Bart Bok in 1936)

c) "It is possible to get around it by supposing ... that stars are constantly being made anew from some diffuse material that consists of nearly pure hydrogen, but the idea is not very attractive." (R. d'Atkinson, also 1936)

d) "The condensation hypothesis now seems suitable only for creation of a few exceptional objects rather than all the stars." (J. L. Greenstein in 1951)

e) "The galactic clusters are a vanishing species that will have disappeared a hundred cosmic years hence." (Bok again, in 1945)

I have letters from Lyman Spitzer (dated 16 November 1996) and Fred Hoyle (10 December 1996) describing their early involvement with ideas of star formation as an on-going process. Spitzer explained that a passage he wrote on the subject in 1939 was regarded by "several astronomers .. as too speculative, and, much to my regret, the paper was published without this introduction". He declined at the time to say who the carpers had been (on the grounds that one of them was still very much alive) and cannot now again be asked. Hoyle, who had briefly supported the then fairly widespread idea that massive stars had been born three Gyr ago at low mass and only just recently accreted most of their material from diffuse nebulae (the origin of Hoyle & Lyttleton's papers on accretion, now generally cited in binary star and high energy astrophysics contexts), explained: "We did not think of on-going star formation because at the time (1939) the astronomical world did not believe in an interstellar medium containing hydrogen ... in 1940 Lyttleton and I postulated the existence of hydrogen of density 10^{-24} g/cm^3. This might have led us to the thought of continuous star formation. But in 1940, with the war just getting going, we had a lot of other problems to cope with." And indeed it was not until the war was over and interstellar neutral hydrogen (at an average density of about 10^{-24} gram or one atom per cubic centimeter) was discovered (Ewen & Purcell 1951) that on-going star formation became a respectable topic for astronomers to work on.

Thus it is that we still hear a slightly tentative tone in CHPG (1952): "Two groups of objects lead to very much smaller ages: the spendthrift stars and the arms of spiral galaxies. As these two phenomena are closely bound up with one another, and the arms of galaxies are made visible and marked out by the stellar spendthrifts, we cannot but conclude that the latter are really young. We can even toy with the idea that stars are being born continuously: and the places where the young stars are found suggest very definitely that they are born of the interstellar dust."

5. THE NATURE OF SPIRAL ARMS AND EARLY EVOLUTION OF GALAXIES

My generation tends to describe spiral arms (at least the elegant grand design ones that go all the way around) as solitons and to attribute the serious work to C. C. Lin and Frank Shu in around 1965, which is not to say that we have necessarily read the papers, much less understood them, with perhaps a nod to Bertil Lindblad of the resonances, who started thinking about what we now call pattern and gas speeds in about 1936. That even he had a precursor, I learned from McVittie & CHPG (1951). His name was Brown (1925). He was at Yale, thought that van Maanan's proper motion showed that spiral nebulae rotate, and considered the possibility of a non-axisymmetric galactic potential.

It turns out that there is an even older tradition of forming spiral arms in an axisymmetric potential, to which the McVittie & CHPG paper belongs. They quote Stephen Alexander (1852!) who was professor of mathematics at the College of New Jersey (now known as Princeton). He had in mind actually making arms by letting something wind up in the rotating disk. Wilczynski (1896,

1899) considered the same situation and, if cited at all, is normally credited with the criticism that spiral arms made this way will wind up out of existence in a few rotations, while spiral patterns are common, and so must be long-lived (or recurrent).

McVittie & CHPG (1951) began with a very simple model, of a dense bar of material (in which stars will form) in an otherwise axisymmetric, differentially rotating disk. They calculated the winding over several rotation periods of about 10^8 years each. A crucial point is that the center winds up first, so that after a billion years or so, the galaxy has a smooth, round central region, a few kpc across, with arms coming out of it. This is actually more or less what we see in many galaxies. McVittie and CHPG did not address what makes the bar, but must have had in mind something that could occur more than once over the life of a galaxy.

The modern descendent of this tradition has been called stochastic self-propagating star formation (SSPSF), is associated with the names of Seiden & Gerola (then at IBM), and is widely blamed for the production of so-called flocculent spirals, in which there are "arm portions" (rather like chicken parts), but nothing that can be traced all the way around. The Milky Way is probably of this type. The idea is that star formation gets started in some small region, propagates outward, and the bright zone of young stars and H II regions is dragged into a curve by differential rotation in the disk. Incidentally, in the McVittie and CHPG picture, the wound-up region presumably includes the galactic bulge. Papers published within the last year have claimed as equally self-evident that the disk formed first and that the bulge is made out of dynamically disturbed disk material.

Finally on this topic, I would like to quote again from "Stars in the Making" (CHPG 1952, p. 125):

"What conclusion is to be drawn from the variety of distributions and motions? I think we can picture the past development of a spiral galaxy in a very general way ... If a mass of gas, originally, spherical, begins to rotate, the gas will probably sink steadily toward the central plane. I do not think that the rate of sinking has been accurately calculated, but I suspect that for a system the size of the galaxy it would be around a thousand million years. Thus we can suppose that the more spherical is the group in which a particular kind of star is arranged, and the more inclined its motions to the central plane, the earlier it was formed in the history of the galaxy. We can picture a galaxy gradually building up its Population II substratum within the steadily sinking mass of dust and gas. The more spherical a subsystem, the older one can suppose it to be."

This sounds, of course, remarkably like the "monolithic" model of galaxy formation and early evolution associated with the names of Eggen, Lynden-Bell & Sandage (1962) a decade later. The missing key idea in the CHPG version is concurrent increase in heavy element abundances during dynamical evolution. One item in the next section may cast some light on this.

6. SOME OF THE REST OF THE REST

It is impossible to look at any part of CHPG's career without coming upon resonances with all sorts of things. This reflects, undoubtedly, both her wide ranging scientific concerns and the "interesting times" in which she lived. For instance, there is a 1938 review of "The Variable Stars" by Barbier (1938, signed only as DB), in which he expresses his "etonnement" that "l'etrange[s]" like the authors do photometry not by the "si pratique and si precise" method of Fabry & Buisson (whatever it was!), but persist in employing extrafocal images of the "Schraffierkassette" to which Gingerich referred in his talk.

Barbier was the next-but-one successor as director at Paris to Mineur, who recalibrated the Cepheid distance scale using Trumpler's (1930) interstellar extinction value (Mineaur 1946) that the authors of "Variable Stars" forgot. Chalonge came in between, and his annual reports from Paris during the war years make interesting reading, with references to "L'application des nouvellse lois raciales" as accounting for the departure of various staff members. Mineur himself was "mis prematurement a la retraile comme un astronome ..." (no reason given). Another staff member "est venu se refugier a l'Institute d'Astrophysique sous une fausse identite". A ha, Schatzman, you will say, if you have read his autobiography in "Annual Reviews". No, actually Jacques Berger (another frequently Jewish name), who does not appear in the post-war membership list of the IAU. Schatzman was a refugee under a false name at Haute Provence and is also mentioned in one of the annual reports.

Normally we associate CHPG with the 1925 demonstration that all stars have pretty much the same chemical composition, for which (unlike the dominance of hydrogen and helium) she quickly received credit. But she later attempted to contribute to the opposite recognition, that all stars do not have the same chemical composition. And this was not well received either. I quote here from a 1995 email from Lawrence H. Aller, her co-author on an unpublished paper:

"Yes, indeed, Cecilia Payne-Gaposchkin and I prepared a paper that was declared anathema by the pundits to whom a copy was sent ... One day, Cecilia was talking to me [LHA was a Harvard postdoc with Menzel after completing his Ph.D. at Lick] about the huge differences in nova spectra during their nebular phase and wondering if this could indicate chemical composition differences - a heresy in those days. I told her I thought we could take a crack at it using the methods of the Harvard Nebular series, assuming that collisional strengths did not differ enormously for different ions of the same electronic configurations. Then you could not reconcile the strengths of the emission lines of N, O, Ne, etc. ions with the same chemical composition in all objects. I allude to this possibility in footnote 45 in ApJ 97, 135, 1943, which is markedly weasel-worded to avoid the wrath of the censors. Cecilia and I sent our paper to Struve & Swings and gave a copy to Menzel. Struve did not bother with it and let Swings do the hatchet job. Later, in 1943, Swings told me that he did not believe that chemical composition [differences] between stars existed! ... Menzel attacked the paper with great emotion, but never did I find out what his actual objections were. At the time I believe that there may have been less than congenial feelings between Menzel and Cecilia. Cecilia decided that in view of the great hostility this paper aroused, it should not be published..."

Not the sort of history that would have led her to look for chemical composition correlations with stellar dynamics.

And then there are the obituaries, with which I especially identify, having just completed four years of being largely responsible for commissioning and editing, or writing, those of deceased AAS members for our Bulletin. Those of Annie J. Cannon (CHPG 1941), "the happiest person I have ever known"), and Otto Struve (CHPG 1963), with whom she felt a good deal of commonality and of Zwicky (CHPG 1974), with whom she did not, especially repay reading (or rereading).

CHPG was, of course, "the first woman ..." for all sorts of things, but she was no all-purpose iconoclast. She mentions in her autobiography that an immigrant should take out citizenship "for the same reason a woman takes her husband's name, out of gratitude." And only moments ago did the following passage (from page 131 of "Stars in the Making" fully register: "... only the incurable optimism of the astronomer can excuse the temerity with which I have described it. The evolution of galaxies is still hot off the griddle; perhaps a more cautious man would have dismissed it as too hot to handle."

ACKNOWLEDGMENTS

It is the traditional perogative of the last speaker at a meeting to express the collective gratitude of the participants to those who have made the gathering possible. Thus we are all grateful to Owen Gingerich, Andrea Dupree, David Latham and the rest of the organizing committee for putting things together so effectively in the short time available and to Irwin Shapiro for institutional support. I personally extend my thanks to Jim Wright, who gave me his personal autographed copy of "The Galactic Novae" some years ago.

REFERENCES

Abell, G. O. 1964 Exploration of the Universe, Holt, Rinehart, Winston, p. 445

Alexander, S. 1852 AJ 2, 95 & 105

Baade, W. & Zwicky, F. 1934 PNAS 20, 254 & 259

Baade, W., Burbidge, E. M., Hoyle, F., Burbidge, G. R., Christy, R. F. & Fowler, W. A. 1956 PASP 68, 296

Barbier, D. 1938 Ann. d'Ap. 1, 430

Bethe, H. 1939 PR 55, 434

Branch, D. & Patchett, B. 1973 MNRAS 161, 71

Brown, E. W. 1925 ApJ 61, 111

Burbidge, E. M., Burbidge, G. R., Fowler, W. A. & Hoyle, F. 1957 RMP 29, 547 (B^2FH)

Eggen, L. J., Lynden-Bell, D. & Sandage, A. 1962 ApJ 136, 748

Ewen, H. & Purcell, E. M. 1951 Nature 168, 356

Gurevich, L. A. & Lebedinsky, A. I. 1947 Dokl. Akad. Nauk USSR 56, 23

Hoyle, F. & Fowler, W. A. 1960 ApJ 132, 565

Hubble, E. P. 1929 PNAS 15, 168

Kirshner, R. P., Oke, J. B., Penston, M. V. & Searle, L. 1973 ApJ 185, 303

Lundmark, K. 1935 Lund Medd. 2, No. 74

McVittie, G. C. & Payne-Gaposchkin, C. 1951 MNRAS 111, 506

Mineur, H. 1946 Ann. d'Ap. 7, 160

Minkowski, R. 1940 PASP 52, 506

Payne-Gaposchkin, C. 1936a ApJ 83, 173

Payne-Gaposchkin, C. 1936b ApJ 83, 245

Payne-Gaposchkin, C. 1936c PNAS 22, 332

Payne-Gaposchkin, C. 1941 Science 93, 443

Payne-Gaposchkin, C. 1952 Stars in the Making, Harvard

Payne-Gaposchkin, C. 1954 Introduction to Astronomy, Prentice Hall, p. 391

Payne-Gaposchkin, C. 1957 The Galactic Novae, North Holland

Payne-Gaposchkin, C. 1963 Sky and Telescope 24, 308

Payne-Gaposchkin, C. 1974 Sky and Telescope 47, 311

Payne-Gaposchkin, C. & Gaposchkin, S. 1938 The Variable Stars, Harvard Observatory Monographs # 5

Payne-Gaposchkin, C. & Whipple, F. L. 1940 PNAS 26, 264

Schatzman, E. 1946 Ann. d'Ap. 9, 199

Trimble, V. 1997 in Star Formation Near and Far, S. S. Holt & L. G. Mundy, eds., AIP Conf. Proc. 393, 15

Trimble, V. 2001 in Young Supernovae Remnants, S. S. Holt et al., AIP Conf. Proc., in press

Trumpler, R. J. 1930 LOB 14, 154

Whipple, F. L. 1939 PNAS 25, 118

Whipple, F. L. 1946 ApJ 104, 1

Whipple, F. L. & Payne-Gaposchkin, C. 1941 Proc. Amer. Phil. Soc. 84, 1

Wilczynski, E. J. 1896 ApJ 4, 97

Wilczynski, E. J. 1899 AJ 20, 6

Zwicky, F. 1936a PNAS 22, 457 & 557

Zwicky, F. 1936b PASP 48, 191

APPENDIX

HCO PINAFORE

It was in 1878 that composer Arthur Sullivan and lyricist William Gilbert had their first truly great success with HMS Pinafore. But they soon began losing so many royalties from pirated copies that the next year, 1879, they came to New York to get matters straightened out. Sullivan was greatly lionized during his American visit, and it was at that time that one of the Harvard College Observatory staff members, Winslow Upton, wrote a parody, the "HCO Pinafore". Among others satirized was Edward C. Pickering, Director of the Observatory, and that may well be the reason that the "HCO Pinafore" was not performed until 1930, when all the characters were safely dead. Under Harlow Shapley's management, the operetta was performed in the director's observatory residence for the American Astronomical Society, and again for the Bond Astronomical Club, with Cecilia Payne singing the lead role of Josephine.

At the banquet of "The Starry Universe" symposium, singers from the Harvard-Radcliffe Gilbert & Sullivan Players, under the direction of Brian Gatten, presented an abridged concert version of the HCO Pinafore. The accompanying commentary was read by Owen Gingerich.

At Harvard College Observatory in the 1870s there were two principal research activities, Pickering's innovative program of photometry, using among other instruments the great 15" refractor, and the traditional effort involving the meridian circle and clocks. The meridian circle was in the wing to the west of the 15" dome, in the space now occupied by Building A. At the far west end of that wing was the computing room, and below it in a lead-lined, constant-temperature dungeon, were the clocks.

The first act is set in the computing room, with a chorus of women computers.

Chorus of Computers

We work from morn 'till night,
For computing is our duty;
We're faithful and polite,
And our record book's a beauty;
With Crelle and Gauss, Chauvenet and Peirce,
We labor hard all day;
We add, subtract, multiply and divide,
And we never have time to play.

No, no; No, no,
We never, never play.
No, no; No, no,
We never, never play.

We sit at our desks all day, all day,
We work from morn 'till night
And computing is our duty,
We are faithful and polite,
And our record book's a beauty,
Computing is our duty,
Our record book's a beauty,
We work from morn 'till night,
We are faithful and polite.

Prof. Arthur Searle, in charge of the photometers, then opens the action.

A Stronomer's Song

I'm called an astronomer, skillful astronomer,
Though I could never tell why;
But yet an astronomer, happy astronomer,

Modest astronomer, I.
I read the thermometers, break the photometers,
Mend them with paper and wax;
I often lament that so seldom is spent
A fair evening on star parallax.
I write many letters, give aid to my betters,
And often sit up late o'nights
To catch a few glimpses of the many eclipses
of Jupiter's bright satellites.

I'm called an astronomer, skillful astronomer,
Though I could never tell why;
But yet an astronomer, happy astronomer,
Modest astronomer, I.

The plot begins to develop when Prof. William Rogers, in charge of the meridian circle, comes in to announce that he fears he is about to lose Josephine, his circle reader, to the gentlemen of Providence who propose to hire her away. Needless to say, Rogers has a crush on Josephine, and he is desperate to think of a plan to foil this raid on HCO. After some good-natured banter about mistakes in the record books, the villain Leonard Waldo and the influential men from Providence arrive, gaily tripping, lightly skipping, gloves and canes and glasses gleaming. Waldo is full of himself, gloating over his rise in reputation since he has left Cambridge for Rhode Island.

"I'm at last an LLD
I'm very proud of my degree
For it shows that I'm a man of extraordinary sense
And so say the influential men of Providence."

At this point Winslow Upton, assistant observer on the photometers, sings of his own ascent. Watch out for a bit of word play here–not only are the photometers coded with letters,but S and P refer to Searle and Pickering:

Winslow Upton's Song

Two years ago I came to be
An assistant at th'Observatree.
I spent my time from day to day
In making computations for the Coast Survey.
My "patent" computations did so well for me
That now I am observing with Photometer P.

Chorus

My "patent" computations did so well for me
That now I am observing with Photometer P.

In the cool night air with "S" and "P"
I wearied my eyes on photometree.
Bright stars with H, faint stars with I
Blue doubles reserved for a cloudy sky.
So many close doubles were observed by me
That now I am observing with Photometer P.

Chorus

So many close doubles were observed by me
That now I am observing with Photometer P.

I pulled the string or turned the screw
    Or drove the match as I was told to do;
I "cranked" the circle near and far
    while "S" strained his eyes to catch the prism star.
I became so cranky they promoted me
    To be an observer with Photometer P.

Chorus

I became so cranky they promoted me
    To be an observer with Photometer P.

I turned the dome with so grand a shock
    that I broke two windows and the Elliott clock;
I burst the gas pipe rolling the chair,
    And created a blaze for the winter's scare.
For my worthy zeal they requested me
    to try my strength on Photometer P.

Chorus

For my worthy zeal they requested me
    To try my strength on Photometer P.

Now Waldo, Wendell, Metcalf, Mann,
    Copy my example as far as you can;
Compute, observe, and – mark my word –
    Your labor will gain its due reward.
And if you're asked "What reward shall it be?"
    Say, "Let me observe with Photometer P".

Chorus

And if you're asked "What reward shall it be?"
    Say, "Let me observe with Photometer P".

The men of Providence have gone off to consult with the Director about hiring Josephine, but they have left a new song, so the chorus of computers along with Professor Rogers, Josephine, and Upton decide to try it:

Astronomer's trio

An astronomer is a sorry soul,
    As free as a caged bird;
His sympathetic ear should be always quick to hear
    The directorial word.
He must open the dome and turn the wheel
    And watch the stars with untiring zeal.
He must toil at night though cold it be
    And he never should expect a decent salaree.

Chorus repeats four lines

His eyes should shine with learned fire,
    His brow with thought be furrowed;
His energetic speech should be ever prompt to teach

The truths which he has borrowed.
His knees should bend and his neck should curl
His back should twist and his face should scowl,
One eye should squint and the other protrude,
And this should be his customary attitude.

Chorus repeats four lines

The news arrives that Josephine has a much better offer from Providence, and the Director is sending her there. Rogers, in despair, hatches a plot to take the faulty prisms out of the photometer, have Josephine take them overnight to Alvan Clark, whose optical shop is about a mile away from HCO, just south of Central Square, and then, by this coup of getting prisms corrected, will be seen by the Director as too valuable an asset to the Observatory to be allowed to leave. Josephine teases Prof. Rogers, but in asides allows that she is really fond of him. But Rogers, thinking he has lost his suit, draws a pistol and threatens suicide. Josephine quickly relents and agrees to take the prisms to Clark as Act 1 ends.

As Act 2 opens in the library, the photometers have been set out for inspection by the visitors from Providence; Pickering, alone, is examining the record sheets of the meridian photometer. This was one of Pickering's inventions, a double telescope with one pointed at his standard, the pole star, and the other bringing the comparison star from elsewhere in the sky into the same field of view. Unknown to Pickering, however, was the fact that Polaris was a Cepheid variable star of low variability–enough to put his magnitudes at sixes and at sevens:

E.C. Pickering's Song

Pole Star, to thee I sing
Bright pivot of the heavens,
Why are all our magnitudes
Either at sixes or at sevens?
I have lived hitherto
Free from the breath of slander,
Beloved by all my crew
A really popular commander.
But now my prisms all rebel
And ruin the photometer,
And damage also, sad to tell,
My fame as an astronomer.
Pole Star, to thee I sing
Bright pivot of the heavens,
Why are all our magnitudes
Either at sixes or at sevens?

Pickering is now called out by Prof. Searle to confer with Mr. Waldo from Providence. Josephine and Prof. Rogers enter, and ask Searle to explain the photometers to them. Searle and Rogers exit, leaving Josephine to capture the prisms:

Josephine's Song

The hours creep on apace,
My guilty heart is quaking;
Oh, that I might retrace,
The step that I am taking!
It's folly, it were easy to be showing,
What I am giving up and whither going.
On the one hand, a new, luxurious home,
With Brussels carpet and no dust or damp,

With wonderful machines to move the dome,
    And all the apartments lit by student lamp –
No inconvenience and no failure, you know,
    For every instrument is made by Grunow.
And on the other, a dark and dingy place,
    all clattered up and smelling strong of oil,
Where record pen most always fail to trace
    and thus rewards the most exacting toil;
Observing sun and moon and many a low star,
    or getting up at midnight for the pole star.
Meridian circle, nobly born,
    so brilliant and well-known,
Which bravely swings from early morn
    till half the night is flown!
No golden rank can it impart,
    no wealth, nor fondest hopes.
No fortune, save its brazen heart,
    and trusty microscopes.
And yet it is so wondrous fair,
    that love for one so passing rare
Were little else than solemn duty.
    Oh god of love and god of reason, say!
Which of you twain shall my poor heart obey?

Waldo enters with Pickering, and it now becomes clear that Waldo is much more interested in Josephine romantically than astronomically, and Pickering congratulates him on the happy result of their negotiation. Waldo leaves, but Upton comes in to break the news that Josephine is stealing the prisms. Pickering hides behind a door as Josephine and Rogers come in to take the prisms, which Josephine promptly drops. Caught in the act, she is sentenced to confinement in the clock room dungeon. Rogers begs in vain to explain while Waldo gloats that "For such wickedness received its fitting recompense." In the nick of time Searle reappears and all is set aright as Pickering declares that Josephine is too valuable to be allowed to leave the institution and the happy couple are reunited as the curtain falls on the HCO Pinafore.

Finale
Chorus:

O joy, o rapture unforseen!
    The clouded sky is now serene
The god of day, the orb of love
    Has hung his ensign high above
The sky is all ablaze
    With wooing words and loving song
We'll chase the lagging hours along
    And if our Josephine is coy
We'll murmur scientific joy
    In dreamy roundelays.

ECP: For I am the captain of this gallant crew.
All: And a right good captain, too.
ECP: You are very, very good, and be it understood
    I command a right good crew
All: We are very, very good, and be it understood
    He commands a right good crew.
ECP: Since moving by my right in society polite
    And among many men of note
I'll never cease to wear with punctilious care

	A tall hat and swallow tailed coat.
All:	What, never?
ECP:	No, never!
All:	What, never?
ECP:	Well, hardly ever!
All:	Hardly ever wear a swallow tail coat.
	Then give three cheers and three times three
	For the gallant captain of the Observatree!

The meeting was held in the Phillips Auditorium, CfA.

Mike Snowden, Bob Kurucz, Owen Gingerich and Ken Janes.

Becky Koopmann and Dorrit Hoffleit.

Virginia Trimble and Paul Hodge.

INDICES

If a page number in the Name Index is in a **bold** font it indicates that the person is the author of a paper. A number in a **bold** font in the Object or Subject Index indicates that the object or subject was mentioned in the title of the paper or is the institution or place of an author. Most subjects were taken from chapter or section headings. An asterisk after a number indicates that that the person or place is in a picture.

OTHER PUBLICATIONS

Name Index

S

T

U

V

W

Y

Z

OBJECT INDEX

SUBJECT INDEX

U

V

W

Other Publications in Astronomy and Mathematics

THE EVOLUTION OF POPULATION II STARS (1972), A. G. Davis Philip, ed., Dudley Observatory Report No. 4.

MULTICOLOR PHOTOMETRY AND THE THEORETICAL HR DIAGRAM (1975), A. G. Davis Philip & D. S. Hayes, eds., Dudley Observatory Report No. 9.

UBV COLOR-MAGNITUDE DIAGRAMS OF GALACTIC GLOBULAR CLUSTERS (1976), A. G. Davis Philip, M. F. Cullen & R. E. White, Dudley Observatory Report No. 11.

AN ANALYSIS OF THE HAUCK-MERMILLIOD CATALOGUE OF HOMOGENEOUS FOUR-COLOR DATA (1976), A. G. Davis Philip, T. M. Miller & L. J. Relyea, Dudley Observatory Report No. 12.

GALACTIC STRUCTURE IN THE DIRECTION OF THE POLAR CAPS (1977), M. F. McCarthy & A. G. Davis Philip, eds., in Highlights of Astronomy, Vol 4, Reidel, Dordrecht.

IN MEMORY OF HENRY NORRIS RUSSELL (1977), A. G. Davis Philip & D. H. DeVorkin, eds., Dudley Observatory Report No. 13.

IAU SYMPOSIUM NO. 80, THE HR DIAGRAM (1978), A. G. Davis Philip & D. S. Hayes, eds., Reidel, Dordrecht.

IAU COLLOQUIUM NO. 47, SPECTRAL CLASSIFICATION OF THE FUTURE (1979), M. F. McCarthy, A. G. Davis Philip & G. V. Coyne, eds., Vatican Observatory.

PROBLEMS OF CALIBRATION OF MULTICOLOR PHOTOMETRIC SYSTEMS (1979), A. G. Davis Philip, ed., Dudley Observatory Report No. 14.

X-RAY SYMPOSIUM 1981 (1981), A. G. Davis Philip, ed., L. Davis Press.

IAU COLLOQUIUM NO. 68, ASTROPHYSICAL PARAMETERS FOR GLOBULAR CLUSTERS (1981), A. G. Davis Philip & D. S. Hayes, eds., L. Davis Press.

A DEEP OBJECTIVE PRISM SURVEY OF THE LARGE MAGELLANIC CLOUD FOR OB AND SUPERGIANT STARS. PART I. (1983), A. G. Davis Philip & N. Sanduleak, L. Davis Press.

IAU COLLOQUIUM NO. 76, THE NEARBY STARS AND THE STELLAR LUMINOSITY FUNCTION (1983), A. G. Davis Philip & A. R. Upgren, eds., L. Davis Press.

IAU SYMPOSIUM NO. 111, CALIBRATION OF FUNDAMENTAL STELLAR QUANTITIES (1985), D. S. Hayes, L. Pasinetti & A. G. Davis Philip, eds., Reidel, Dordrecht.

IAU COLLOQUIUM NO. 88, STELLAR RADIAL VELOCITIES (1985), A. G. Davis Philip & D. W. Latham, eds., L. Davis Press.

HORIZONTAL-BRANCH AND UV-BRIGHT STARS (1985), A. G. Davis Philip, ed., L. Davis Press.

SPECTROSCOPIC AND PHOTOMETRIC CLASSIFICATION OF POPULATION II STARS (1986), A. G. Davis Philip, ed., L. Davis Press.

STANDARD STARS (1986), A. G. Davis Philip, ed., in Highlights of Astronomy, Vol 7.

IAU COLLOQUIUM NO. 95, THE SECOND CONFERENCE ON FAINT BLUE STARS (1987), A. G. Davis Philip, D. S. Hayes & J. W. Liebert, eds., L. Davis Press.

IAU SYMPOSIUM NO. 126, GLOBULAR CLUSTER SYSTEMS IN GALAXIES (1988), J. E. Grindlay & A. G. Davis Philip, eds., Reidel, Dordrecht.

NEW DIRECTIONS IN SPECTROPHOTOMETRY (1988), A. G. Davis Philip, D. S. Hayes & S. J. Adelman, eds, L. Davis Press.

CALIBRATION OF STELLAR AGES (1988), A. G. Davis Philip, ed., L. Davis Press.

STAR CATALOGUES: A CENTENNIAL TRIBUTE TO A. N. VYSSOTSKY (1989), A. G. Davis Philip & A. R. Upgren, eds., L. Davis Press.

THE GRAVITATIONAL FORCE PERPENDICULAR TO THE GALACTIC PLANE (1989), A. G. Davis Philip & P. K. Lu, eds., L. Davis Press.

CCDs IN ASTRONOMY. II. NEW METHODS AND APPLICATIONS OF CCD TECHNOLOGY (1990) A. G. Davis Philip, D. S. Hayes & S. J. Adelman, eds., L. Davis Press.

PRECISION PHOTOMETRY: ASTROPHYSICS OF THE GALAXY (1991), A. G. Davis Philip, A. R. Upgren & K. A. Janes, eds., L. Davis Press.

Mm, Vol I, SERIES ON FRACTALS, MIDGETS ON THE SPIKE (1991), A. G. Davis Philip, A. Robucci, M. Frame & K. W. Philip, L. Davis Press.

OBJECTIVE-PRISM AND OTHER SURVEYS. A MEETING IN MEMORY OF NICHOLAS SANDULEAK (1991), A. G. Davis Philip & A. R. Upgren, eds., L. Davis Press.

NEW YORK STATE ASTRONOMY (1992), A. G. Davis Philip, ed., L. Davis Press.

WORKSHOP ON DATABASES FOR GALACTIC STRUCTURE (1993), A. G. Davis Philip, B. Hauck & A. R. Upgren, eds., L. Davis Press.

IAU SYMPOSIUM No. 167, NEW DEVELOPMENTS IN ARRAY TECHNOLOGY AND APPLICATIONS (1995), A. G. Davis Philip, K. A. Janes & A. R. Upgren, eds., Kluwer Academic Publishers, Dordrecht.

PHOTOMETRIC SYSTEMS AND STANDARD STARS (1996), V. Straižys & A. G. Davis Philip, eds., Baltic Astronomy 5 and L. Davis Press

YALE PARALLAX CATALOGUE (1996), William van Altena, John Lee & Dorrit Hoffleit, L. Davis Press

30 YEARS AT THE VAN VLECK OBSERVATORY (1996), A. G. Davis Philip, ed., Baltic Astronomy 6 and L. Davis Press

THE THIRD CONFERENCE ON FAINT BLUE STARS (1997), A. G. Davis Philip, James W. Liebert & Rex A. Saffer, eds., L. Davis Press

ELECTRONIC PUBLISHING: NOW AND THE FUTURE (1998), Peter B. Boyce & A. G. Davis Philip, in Highlights of Astronomy, Vol. 11A, J. Andersen, ed., Kluwer Academic Publishers, Dordrecht, p. 595

ANNI MIRABILES: A SYMPOSIUM CELEBRATING THE 90th BIRTHDAY OF DORRIT HOFFLEIT (1999), A. G. Davis Philip, William van Altena & Arthur R. Upgren, eds., L. Davis Press

THE Kth REUNION (2000), A. G. Davis Philip, ed., L. Davis Press